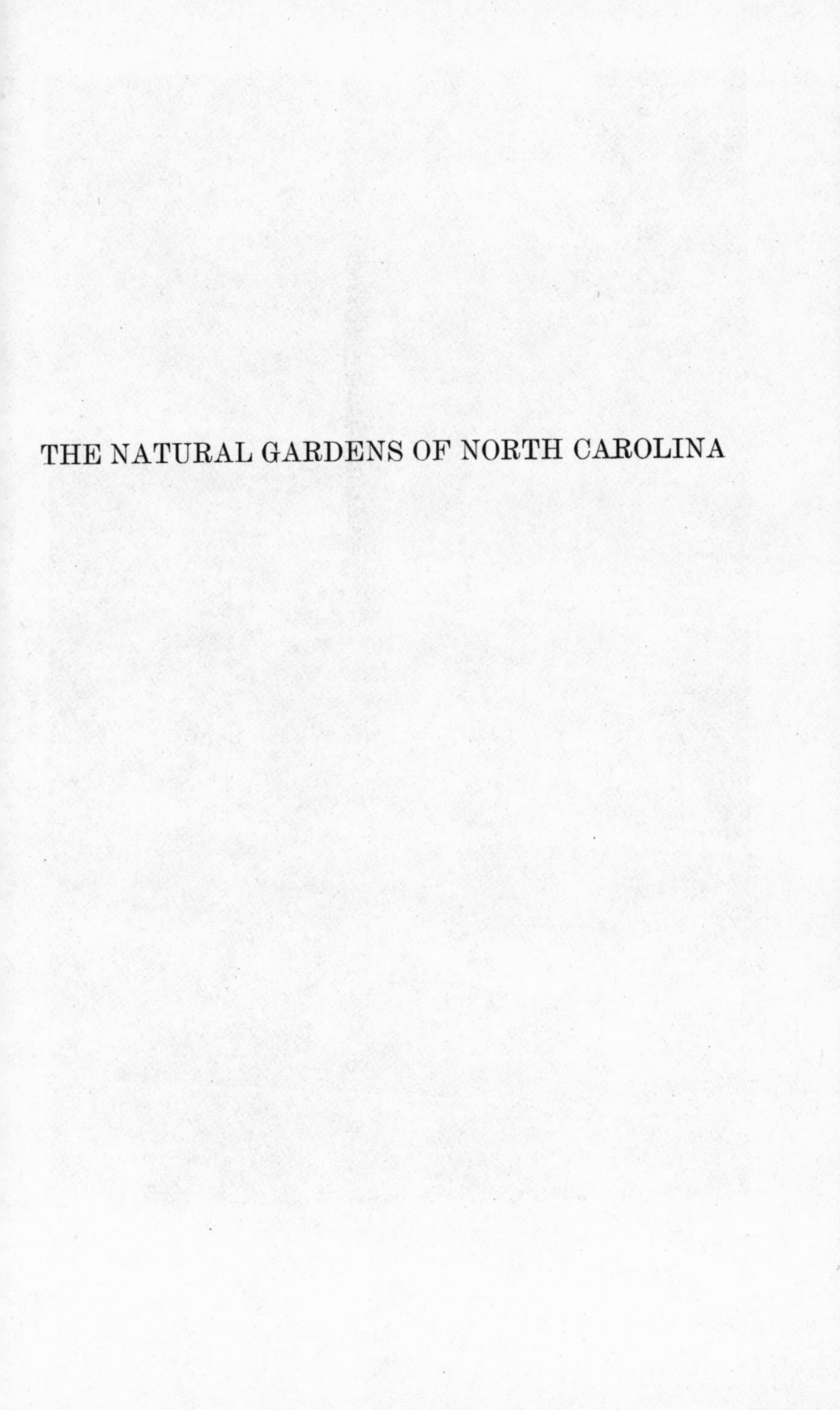

THE NATURAL GARDENS OF NORTH CAROLINA

THE GREAT FOREST NATURAL GARDEN REACHES ITS FINEST DEVELOPMENT ON MOIST COOL MOUNTAIN SLOPES. LINVILLE GORGE NEAR THE FALLS. PHOTO THROUGH COURTESY OF WOOTEN-MOULTON

THE NATURAL GARDENS OF NORTH CAROLINA

with

Keys and Descriptions of the Herbaceous Wild Flowers Found Therein

by

B. W. Wells

The University of North Carolina Press

CHAPEL HILL

Manufactured in the United States of America
Library of Congress Catalog Card Number 68-15215
ISBN 0-8078-1058-4

DEDICATED TO THE MEMBERS

OF

THE GARDEN CLUB OF NORTH CAROLINA

PREFACE

In recent years a new emphasis has appeared in the field of plant study, which involves the attempt to understand the plant in relation to its environment. It tries to answer such an important question as why plants grow where they do and the equally significant one of why they are not present when absent from an area. Organism and environment constitute the real whole, so that ecology, the science which deals with both in their relation to each other, is becoming increasingly valuable as a major science helping us to understand the world about us. In this study the ecological approach is emphasized.

The first part of this popular book dealing with our natural gardens is devoted to a general account of the vegetation and habitat of each of the eleven major plant communities of North Carolina. In the second part, an original artificial key to the herbaceous wild-flower plants of the state, is presented accompanied by description of the genera and important species. Though fully aware of the difficulties involved, the author has initiated the key on the ecological basis previously mentioned. Thus, all of the savannah or grass-sedge bog wild flowers, for example, are treated together, and the fresh water marsh species, in another place. Approximately two-thirds of the wild flowers fall into the "great forest" group, which involves most of the state since the well-drained upland habitat is determining here. For key purposes the eleven major communities have been condensed to seven. Plants definitely ranging into more than one major habitat have been included in both, or, in some instances, three "homes" are involved.

The writer has always believed that a popular account of vegetation systematically considered, should be an introduction to the professional manuals and not a mere repetition of these. Hence technicalities of all kinds have been avoided so far as possible. The emphasis is upon the genus or group of related species or kinds of plants. Only the professional botanist is interested in recognizing the forty species of goldenrod found in our area or the sixty species of aster. Even when the species are much fewer in number and separated by relatively unimportant characters, the author has stopped at the genus, realizing that such a complete differentiation of species should have no place in a popular account. It has been our aim to assist the amateur plant lover to learn the common names of the herbaceous wild flowers of North Carolina, most of which are recognized by a genus common name.

The woody plants of the state have not been included. Coker and Totten's recently revised *Trees of North Carolina* covers that field in an excellent manner. The shrubs have been described in Apgar's *Shrubs of North America* and in other manuals. Wild flowers in the popular mind are definitely the herbaceous plants; hence these, with a few exceptions, and these only, are dealt with in this account. It may be of interest to state here that of the latter, some five hundred genera are described in this book.

PREFACE

The author has freely used the standard technical manuals as source books: Britton and Brown's *Illustrated Flora of the Northern States and Canada,* Gray's *New Manual of Botany*, and especially Small's *Flora of the Southeastern United States.* The nomenclature and sequence of the plant groups of the latter work have been adopted in this book. It is to be hoped that through this introductory work, many will be encouraged to purchase and use these larger professional manuals.

The appearance of this volume at this time has been made possible through the cooperation of The Garden Club of North Carolina and the University of North Carolina Press. Among the Club officials, mention must certainly be made of the very practical and enthusiastic support of the project by Mrs. S. H. Tomlinson of High Point, and to Miss Susan Iden of Raleigh goes the credit of originating the idea of The Garden Club sponsoring such a book as the one in hand.

Dr. I. V. Shunk, colleague of the writer, and Miss Beulah Weathers, secretary, have assisted materially in checking the many details connected with the botanical features. For this aid the author expresses full appreciation.

B. W. WELLS.

Raleigh, N. C.
November 14, 1932.

TABLE OF CONTENTS

PART II

THE HERBACEOUS WILD FLOWERS OF THE NATURAL GARDENS

INTRODUCTION

North Carolina is unique among the eastern states for possessing within her borders the best examples of the most diverse vegetations as these two criteria are judged in combination. Whoever the men were who designed the geographical biscuit cutter which sliced out the Old North State, they succeeded so well botanically that one might think of them as possessed with less political sense than vegetational acumen. In one east-west state unit they succeeded in including the very finest examples of the southern Appalachian high mountain plant communities, which constitute the southern extension of the Canadian balsam fir forest, along with very extensive developments of typical southern low country plant associations, savannahs, pocosins and swamps which range northward from the Gulf. In a very real sense North Carolina, though lying at right angles to the north-south longitudinal lines, unites Canada and Florida within a little over two-thirds of her length. On the same winter days when sub-zero weather and deep snows are holding the Christmas tree forest of balsams and spruces in a death-like silence, the palmetto trees of Smith Island are softly vocal with the summer-like whisperings of warm breezes fresh from the Gulf Stream. John Brickell, early North Carolina naturalist, was correct when he wrote in 1737, "Of the Plants growing in this Country, I have given an Account of not the hundredth Part of what remains; a Catalog of which would be a Work of many years, and more than the Age of one Man to perfect, or bring into regular classes, this Country being so very large and different in its situation and its Soil.''

And today with but ten million acres under cultivation out of the state's thirty-one million two hundred thousand acres, North Carolina is still an area of vast natural resources in a wild condition, and we use this phrase, ''natural resources,'' not only in the sense of monetary value but in that of educational and aesthetic resources as well—the kind of value implied in our title ''Natural Gardens.''

The vegetation of North Carolina, approached from the ecological point of view, falls into eleven major plant communities or distinctive types of vegetation associated with distinctive habitats. Beginning, for convenience, with those near the sea these natural gardens are: (1) The dune or upland seaside community, (2) salt marsh, (3) fresh water marsh, (4) swamp forest, (5) aquatic vegetation, (6) shrub-bog or pocosin, (7) grass-sedge bog or savannah, (8) sandhill, (9) old field community, (10) great forest, (11) boreal or high mountain forest.

Close observation within these various communities discloses the possibility of a further classification into sub-communities. And these, in turn, may be correlated with slight changes in the habitat factor complex. However, these minor contrasts in the vegetation and habitat are at the present time only of interest to the specialist in ecology, and it must be confessed he understands them none too well.

The attention of the reader must be directed to the

fact that in many places the vegetation seems "all mixed up," and difficult to classify ecologically. Such areas are localities where destruction of the earlier stabilized vegetation has occurred and a process of plant succession is under way. Descriptions of such rapidly changing communities, which may involve the transition from one major community to another, will be given in connection with the various principal plant associations. Such apparently confusing areas yield readily to the understanding with regard to major features after a reasonable amount of ecological field experience.

In the succeeding brief chapters on the eleven major communities of plants, it is hoped that sufficient material has been presented to assist the plant lover measurably in his or her progress in the process of thinking ecologically. This way of thinking, let it be again asserted, is synonymous with thinking fully and inclusively, and only when we are thinking in this manner do we make advances in the pursuit of truth and genuine understanding.

ADDENDUM

In the reprinting of this book no change has been made except the elimination of a few paragraphs from the Preface and Introduction.

Since in relation to three of the major communities, some changes are desirable based upon important, newly discovered data regarding environmental controls, these are briefly presented.

The Seaside Plant Community (Chapter I) In the spring of 1937 the author and Dr. I. V. Shunk found, near Fort Fisher, the numerous tender shoots developing over the sloping tops of the seaside shrubs, blackened and curled. From a Wilmington druggist we obtained distilled water and a bottle of 3 per cent salt solution with which we sprayed unaffected shoots on the lee side of the shrubs. The following day we found those treated with the salt equivalent of sea water wilted and turning dark. Back from the sea out of the spray zone, the shrubs were normal in shape, though still fully exposed to the wind. Later studies, especially the monograph on "The Salt Spray Community" by Stephen G. Boyce, a North Carolina State University graduate student, fully confirmed the original observations.

Both in the text (p. 13) and under the illustration (p. 19) "wind" should be replaced by "salt spray."

The Boreal Forest Community (Chapter X) In this chapter the high mountain grass balds are discussed with no definite conclusions reached as to their origin. Later, in hiking over the well-known Indian trail on the ridge top of the Balsam mountains, I noted that extensive areas were covered with the oatgrass and wild flowers of the balds indicating these to be but expanded trails of the early hunter Indian. Climatic factors may be ruled out since the typical bald boundary changes from grass to forest within a few feet. Most important was the dying tree seedlings as they failed in competition with the heavy grass cover. More significant data is given to support the Indian origin theory in my article "Southern Appalachian Grass Balds" (*The Journal of the Elisha Mitchell Scientific Society*, Vol. 53 [1937], pp. 1-26).

The Savannah Community (Chapter VI) Of all the communities of this type the large one near Burgaw in Pender County was pre-eminent. It was given intensive study during two summers, the results of which were published in Technical Bulletin 32, Agricultural Experiment Station (1928).

This 1,500 acre area was probably the most beautiful wild flower garden in the eastern United States. An attempt was made to have it become a state park and a tourist attraction. Officers of the State Garden Club and the author made an appeal to a legislative finance committee to purchase the area. The tourist potential at that time had not been recognized and the fact that to maintain it in prime condition involved burning over the area every winter were against the project, even though herbaceous fires may easily be controlled.

The discovery of how to farm this land, long regarded as nonagricultural, was made by a truck grower. He dragged the deeply plowed soil into ridges high enough to keep the roots alive on this normally poorly drained area. Recently (summer of 1967) the larger half, 800 acres, was found planted in corn, making it one of the largest corn fields in the state.

The Natural Area Council, a national organization devoted to the preservation of undisturbed native vegetation, has been searching for a runner-up to the Burgaw savannah. In their inquiry the president, R. H. Pough, writes: "The Big Savannah near Burgaw is now gone and nothing can bring it back. Let us at least salvage from this disaster the determination to find several next best savannahs and see that they are given protection in time."

B. W. W.

September 1967

PART I

THE NATURAL GARDENS OF NORTH CAROLINA

CHAPTER I

SEASIDE PLANT COMMUNITIES

The Windy Dunes

The influence of the sea upon every shore line is a very profound one. Especially in the sorting action of moving water with its resulting accumulation of sand, we have one of the most pronounced effects of the ocean's never ending activity. In strong contrast to a rock-bound shore like that of Maine, the geological factors in our coastal region have given us a practically unbroken shore of shifting sand. Only in one very restricted spot near Fort Fisher is there an outcrop of rock. And this rock, it is of interest to note, is made entirely of sea shells.

To a student of plant distribution, an area naturally devoid of vegetation is always of interest, since such areas are always exceptions to the rule. The seaside presents us with two of these soils which cannot become naturally populated with a vegetation. These areas are both characterized by shifting sand; in the one case, the ocean strand (Fig. 1), the particles are moved by water, and in the other, the moving dune (Fig. 2), by wind. Plants have no chance where the soil is shifted about.

With regard to the plant life by the sea, the dominant herbaceous plants are almost strictly confined to the smaller accumulations of sand and, strangely

FIG. 1. AN UNUSUAL SOIL IN NATURE, A PLANTLESS AREA, THE OCEAN STRAND. WITH EVERY TIDE THE SAND IS DISTURBED MAKING ESTABLISHMENT OF THE SALT PLANTS IMPOSSIBLE.

FIG. 2. WHERE SAND "DROWNS" THE FOREST: A MOVING DUNE ON ROANOKE ISLAND.

FIG. 3. DUNE INFANTS ON THE TIDAL FLAT. WHEREVER A BIT OF GRASS BREAKS THE WIND, THE SAND PILES UP.

enough, are never seen inland. The habitat is unusual, not only in its soil of pure sand, but also in certain other factors such as the prevailing shoreward winds, the salt spray from the ocean, and the reflection of heat and light from the light colored substratum.

Perhaps the best introduction to the dune vegetation is the story of the dune's origin, which involves the dune grass as an important part of the tale. The chapters in this story may be observed in innumerable places along our coast, especially on the dune area southward from Cape Hatteras, where the shoreward winds are not so strong.

On the sand-flat just above high tide, a seedling sea oats *(Uniola paniculata)* plant becomes established. When its leaves extend above the surface, a small windbrake is formed, with the immediate result that a

FIG. 4. YOUNG DUNES. THE GRASS AND DUNE GROW UP TOGETHER.

tiny mound of sand accumulates behind and against the little leaf cluster (Fig. 3). This additional bit of soil makes possible further growth of the grass, which improves the windbrake; more sand piles up, more grass grows (Fig. 4), and in a few years a dune many feet high is built. The underground stems and roots of the grass permeate the dune, making a veritable hidden net in which the sand is literally caught and held against the wind.

This captor grass, the sea oats, is by far the most important dune grass on our coast (Fig. 5). The more northerly sand grass *(Ammophila)* enters North Carolina but is seldom seen on the southern half of the state's coast line. And it is well that this is so, since the northern grass cannot compare with the sea oats in its decorative value. In August, when the

FIG. 5. THE STATELY SEA-OATS GRASS FOREVER WAVES IN THE PERENNIAL WIND OF THE SEASIDE.

ample panicles of the large spikelets are fully mature, the dunes are aglow with immense numbers of golden plumes waving against the azure skies of late summer.

On the coast north of Cape Hatteras are found the huge, bare dunes which form under conditions of stronger wind. As we have already indicated, sand is moved too readily and too frequently for plant life to become established; hence the ocean shore on the north-east-facing beaches is low and flat except where the sand has been rolled on to these great barchanes, as such large dunes are called. So rapidly does the high wind move the sand up the slopes and over on the other side that not only can no plant life settle down here, but, like the rolling stone gathering no moss, these bare dunes move slowly across the land. Such dunes

FIG. 6. CONSPICUOUS ON OUR DUNES IS THE SHRUBBY SEA-KALE.

are often a quarter of a mile long and eighty to ninety feet in height.

It was from one of these dunes near Kitty Hawk, North Carolina, that the Wright brothers carried out their experiments with gliders. To make the monument on this dune permanent, it was necessary to plant sand grasses and shrubs extensively and by so doing to assist nature in capturing this loose sand mass.

The dominant plants of the inhabited dunes are so few that they can be dealt with individually in this description of dune vegetation. The most important pioneer, the sea oats, has already been mentioned. Next to it but occurring only in scattered bright green clumps is the sea elder *(Iva imbricata)*, a low shrub 2-3 feet in height (Fig. 6). Similar to it in aspect is the sea kale *(Cakile sp.)*. This shrub has curious fleshy leaves from which water may be squeezed, a character possessed by many desert plants. When it is in fruit it may be readily distinguished by the number of pod joints, the sea elder having but one unit while the sea kale has two. With regard to the latter, it is of especial interest that one of these joints after separation will float while the other sinks. In this way the plant apparently is taking care of keeping up the home population as well as engaging in emigration. Occurring on the undulating surface of the dunes, along with those plants already mentioned, is the gray shrubby croton *(Croton punctatus)*, easily recognizable by its whitish-gray, oval leaves and insignificant flowers. Crush the leaf and a strong fragrant odor is detectable. The occasional occurrence of colonies of bear-grass *(Yucca spp.)* is also to be noted.

One little plant always to be expected is the prostrate dune spurge *(Euphorbia polygonifolia)*. It is an annual, yet successfully maintains itself. The branches radiate from a common root and bear short, narrow leaves. The milky juice will make certain the identification.

As might be expected, certain weeds can tolerate the dry, sterile conditions of the dunes. Some of the more frequently seen are the salt wort *(Salsola Kali)*, a variety of the Russian-thistle of the northwest. A species of pigweed *(Amaranthus pumilus)* and the cocklebur *(Xanthium echinatum)* are not uncommon. The horse-weed *(Leptilon)* in a repressed or dwarf form (Fig. 206) is also evident on most coastal sand areas, particularly on the flats behind the dunes.

On the landward side of the low dunes and ranging

into the sound margin where that is present, is the short spartina or salt grass *(S. patens)*. This sparse, wiry grass is very characteristic of the low flats behind the dunes but is never seen on the sea-front dunes. Sea-beach grass *(Panicum amarum)* is also frequent. Along with the upland salt grass will be found the sea-side evening primrose *(Oenothera humifusa)* and the dune ground cherry *(Physalis viscosa)*. Other scattered herbs found on our dunes are included in the key to the dune wild flowers.

Many people imagine that the water below the dunes is salty since the ocean and often the sound as well are very near. Investigation has shown, however, that the rainfall of the region is sufficiently high to keep the underground salt water pushed back, so to speak, and thus maintain fresh soil water. This will be true on very narrow banks where dunes and dune plants occur.

The drouth-adapted nature of the dune plants, unlike that of the salt marsh, is not related to a salt-water substratum but is due to their exposure to the ever-present wind plus the inability of water to move rapidly or far by capillarity in the dune sand. On dry, hot days one may observe the sea oats' leaves curling tightly, which means that the longitudinal ridges on the upper side of the leaf are pressed together, checking materially the loss of water vapor through the leaf pores. Whenever the stomata or leaf pores open into a closed space, the diffusion of the water vapor into this quiet air is much slower than it would be where the air is moving by the pore opening. Hence such devices as the inrolled ridge-and-groove leaf found in

FIG. 7. FOREST AISLES AMID THE LIVE OAKS.

dune, salt marsh, and prairie grasses, are very effective in water conservation.

Growing in the sand in and about sea-coast towns, one may meet a number of foreign plants which arrived there in the early days when ships dumped ballast earth ashore. Growing like a weed along the street sides in Southport, is a western gaillardia *(G. lanceolata)*, together with the less common spiny Mexican poppy *(Argemone mexicana)*. Species of yucca are also common.

THE SEASIDE FOREST

Just as the herbaceous vegetation on the sea front is peculiar and distinctive, so the forest in North Carolina, when it comes down to the sea, changes its complexion. Every inlander who visits the beaches will be sure to hear about the remarkable "dead live oaks" of

FIG. 8. SUNSHINE AND WIND MAKE THEIR IMPRESS ON THE LOW COMPACT FOREST OF LIVE OAKS. A VIEW OF SMITH ISLAND FROM THE CAPE FEAR LIGHTHOUSE.

the vicinity, if he doesn't see them. It is to be hoped he may not so much see the dead as the live ones for the live oak *(Quercus Virginiana)* is one of our most beautiful trees, with its small hard, gray-green and evergreen leaves borne thickly over a low, rounded head. The trunk and branches are so twisted and bent that "crooked as a live oak limb" is a byword among certain coast folk (Fig. 7). This character in a tree, however, gives it a high place artistically, and it may be truly said of our live oaks that no other trees can compete with them in the beauty of their rustic, scraggly trunks.

Perhaps the best example of this type of forest is found on Smith Island, that little triangular mass of land lying at the mouth of the Cape Fear River, where

FIG. 9. CHAINS OF LITTLE STARS MAKE UP THE PALMETTO INFLORESCENCE.

it marks the point of a great angle in the coast line and thus is wind-swept at all seasons. The forest here is practically a pure stand of live oaks (Fig. 8), with dogwood, red bay, red cedar, yaupon, and wild olive scattered as subdominants beneath the trees. Of especial interest, however, are the sabal palmettoes which have journeyed up the coast from Florida in ages gone by, and are still holding their own amid the oaks, with their bushy heads of fan-like leaves held oftentimes proudly above the surrounding dominant trees. In flower the palmetto presents a starry aspect (Fig. 9). Here, as further south, the variation in the trunks may be observed—some with the leaf stalks hanging on, while other like bald-headed men soon loose their covering. While not abundant, they nevertheless give the island a distinctly tropical aspect and stand as the best

FIG. 10. THE LONE SURVIVOR: A PALMETTO TREE ON SMITH ISLAND NOT KILLED BY THE SHIFTING SAND.

FIG. 11. THE REAR GUARD: PALMETTOES ON SMITH ISLAND RELUCTANTLY GIVING WAY BEFORE THE ADVANCING SEA. PHOTO BY R. F. GRIGGS.

indicator of the proximity of the warm Gulf Stream not many miles offshore.

The seaside woodland, unlike the inland forest, is in many places in danger from two unusual enemies—sand and sea. Where the dune herbs are not in full control the sand mass will be moved slowly under the wind, the particles rolling up the windward side and down the leeward (Fig. 2). Extensive forests in many places have been destroyed in this way. Occasionally a tree escapes where the sand never became too deep as it passed (Fig. 10). At many points the sea is vigorously attacking the land, the strong shore currents sweeping the soil away and, with it, the forest. On Smith Island this kind of forest destruction has been going on for many years. The last survivors of the trees here are the palmettoes (Fig. 11).

Of the minor woody plants of the coast, the very beautiful yaupon shrub, the wild olive, and the pilentary tree deserve especial mention. The first may be readily recognized by its small, shiny, box-like leaves and, when in fruit, by its clusters of red berries. It is our only species of *Ilex* or holly which is used for making tea—and a very satisfying hot drink it is which can be made from the dried young leaves. The Indians were well aware of its value, only they were not content to use it moderately and normally but at their feast times they would make such a strong concoction from these leaves that it would induce immediate vomiting, a desirable happening since that would enable them to dine again without delay. The species name "vomitoria" records this curious custom. The wild olive *(Osmanthus Americanus)* may be identified by its graceful, smooth, opposite leaves, tapering at each end.

FIG. 12. THE CURIOUS PILENTARY TREE WITH ITS SPINE-TIPPED CORKY PROMINENCES.

Systematically it occurs in the same family with privet and lilac. The pilentary tree *(Xanthoxylum Clava-Herculis),* which is found more frequently in open and exposed places, may be instantly known by its sharp spines, which are not only found upon the leaves, but which in a most defensive manner project from corky cushions on the trunk (Fig. 12). Externally curious as it is, one's interest will be more successfully aroused when a piece of freshly cut bark is placed upon the tongue. A most surprising illusion of ice-like cooling is experienced.

Up on that part of the coast north of Hatteras which faces toward the northeast and which is under the direct impact of the well known "nor' easters" the live oak trees which are isolated and fully exposed do

FIG. 13. THE TREE THAT GREW LYING DOWN; AN ISOLATED LIVE OAK GROWING SIDEWAYS UNDER THE INFLUENCE OF THE WIND.

not grow upward but sideways (Fig. 13). This is due to the killing of the branches on the windward side so that only those on the leeward side may go on developing. After years of difficult growth the tree's crown is many feet, horizontally, from the trunk base. The tree's life seems to hang by a narrow margin, for it is always half dying and half living. As a perennial fight against adverse circumstances such a lonely semi-dying tree stands a silent sermon on the text "Never give up."

The woody plants just described stand in a successional relation to the herbaceous species. As the dunes grow larger and less exposed on the landward side, the yaupon and the live oak will appear. The latter will in time assume dominance and maintain its integrity as a forest community for an indefinite time. On the

FIG. 14. LIKE A PRAIRIE, THE GREAT SALT MARSH ON THE SOUTH SIDE OF ROANOKE ISLAND, IS AN IMPRESSIVE LANDSCAPE. NOTE THE FINE TEXTURE OF THE VEGETATION DUE TO THE PREDOMINANCE OF NARROW LEAVES.

mainland near Fort Caswell south of Southport, as the observer goes seaward, he may see the inland forest of turkey oak and long leaf pine community gradually become transformed into the live oak and yaupon forest of the dune line where it had long ago supplanted the dune grass and sea kale.

In summary, let us emphasize again the striking change in the vegetation as the ocean is approached. And let it be confessed that the ecologist has yet much to learn in the way of explanation of this remarkable change in the vegetation, as one goes down to sea.

THE SALT MARSHES

North Carolina's coast is rich (or shall we say poor) in salt marshes. If one has something of the tastes of the naturalist or an artist he will feel very sure that these vast salt-water grass and rush-covered areas have much wealth in them, whereas to the economist's mind they are but waste land. Though they do not contain a large number of species, like the dune vegetation, the few species which are found are of much interest because of their special adaptations to the habitat.

The large extent of our salt marshes (Fig. 14) is, of course, related to the fact of the great development of sounds in North Carolina, the greatest on the eastern seaboard. In these sounds, wherever the water is shallow enough at high tide, the salt grass has established itself. The unusual beauty of the sound shores is due largely to the broad, flat, bright yellow-green masses of the salt grass contrasting with the blue water.

Even the casual observer will note that the vegetation along the landward marsh border occurs in well marked zones correlated with the depth of the water. In the deeper places and constituting the body of the marsh are two or three species of the Spartina grass, a group of grasses related to certain western prairie species which are adapted to survive long drouth periods. Though they are rooted in the salt water soaked soil, these plants are little better off than if they were growing in dry soil. The reason for this will be explained later in the discussion of the salt marsh habitat.

The inner and shallower margin of our marshes almost everywhere carries a heavy growth of the salt marsh rush *(Juncus Roemerianus)*. The leaves of this plant are giant green needles which stand in tussocks and masses 4-5 feet high. In old areas of thickly ag-

FIG. 15. OF A DELICATE ROSE TINT, THE SALT MARSH SABBATIA WITH ITS NARROW LEAVES IS ONE OF THE MOST ATTRACTIVE WILD FLOWERS AMID THE SALT GRASS.

gregated plants this type of marsh is difficult to penetrate. Next to the rush zone is found the transition to the dry bordering sands. In this moist soil with brackish water are found a number of grasses and sedges. One of the most frequently seen grasses is the tall panic grass *(Panicum virgatum)* with its peculiar scaly underground stems. In this zone occur also a number of attractive wild flowers such as the beautiful marsh morning glory *(Ipomoea speciosa)*, the sabbatia (*Sabbatia calycina*, fig. 15), and the marsh aster *(Aster tenuifolius)*. Especially prominent in this transition area are the shrubby water bush *(Baccharis)*, and the salt marsh mallow *(Kosteletzkya)*.

Occurring in patches of varying size in the salt-grass quilt are some plants of especial interest (Fig.

FIG. 16. THE SALT MARSH GOES IN FOR NATURAL GARDENS. SLIGHT CHANGES IN WATER DEPTH GENERALLY ACCOUNT FOR CHANGES IN THE VEGETATION.

16). Few who have been to the beach, will have missed the glasswort or samphire *(Salicornia sp.)*, especially if their visit was in the fall when these curious leafless plants are decorated with brilliant red colors. The erect, fleshy, jointed stem, green at first but changing to the warm colors, are so distinctive that everyone desires to know its name. Another common society of one species is the sea ox-eye *(Borrichia frutescens)*. This is a low shrub which spreads over extensive areas competing with the marsh grass. It has thickish leaves and heads like a small sunflower. (An especially interesting plant is the sea lavender *(Limonium)* which has a branching flower stem arising from the water or mud, bearing small, delicate lavender blossoms scattered along the branches. Each flower base is enclosed

FIG. 17. THE SALT MARSH LOOSESTRIFE EXHIBITS THE REDUCED AND NARROW-LEAVES SO CHARACTERISTIC OF SALT WATER PLANTS.

in a small bract. These plants are more or less scattered, seldom forming pure societies. On tidal flats where little vegetation has become established are the spreading mats of the sea purslane (*Sesuvium maritimum,* fig. 153).

With regard to the problem of salt water soaked soil in which these plants live, it may be strikingly expressed in the old saying about the shipwrecked sailor, "Water, water everywhere, nor any drop to drink." Every schoolboy in this day of science knows that if you fill a membranous bag with salt water or sea water and place it in a vessel of fresh water, in a few hours the bag will, through the process called osmosis, take up so much additional water as to burst it. In the case of the salt marsh plants it is readily seen that with so much salt in the liquid around the root the plant may have difficulty in taking in water against the pull, so to speak, of the large amount of dissolved materials in the sea water. So these plants, like those of prairies and deserts which also have a low water supply, show in their leaf structure various adaptations for checking the outgo of water. First of all they have relatively small leaves (Fig. 17). The blades of the grasses have ridges on the upper side which, when the leaf curls, close together along the tops and thus introduce a most successful check to evaporation. Many of the plants have fleshy leaves or stems as in the glasswort, which character is, as in the case of the cactus, an adaptation to low water supply.

The situation for the salt marsh plants is aggravated by their proximity to the persistent sea winds, for in the summer the almost daily landward breeze of warm air exerts a marked drying effect.

Though these plants are entirely capable of tolerating the salt water, they cannot become established where the water is moving too rapidly as it is near the mouths of inlets or on the ocean front itself. Wherever the sand is moved about too freely little or nothing can become rooted. Hence it is that in the protected and quiet waters of the sounds the salt grass and its associates may develop their peculiar community undisturbed.

Another feature to be mentioned is the extraordinarily tight sod developed by the roots of the salt grass. This compact mass tends to become thicker, and eventually, if water level conditions are favorable, it will go over into peat of a characteristic type recognizable as salt marsh peat. In areas that are now many miles from the coast and that have become fresh water in nature, the salt marsh peat underlies later and more recent accumulations, telling the story of tidal flats of long ago.

CHAPTER II

INLAND MARSHES

Cattails and Their Neighbors

In this community may be seen what nature can do with mud. Bogs in the rainy periods are black and soft enough but in the drouth seasons they become sufficiently dry to be visited comfortably in ordinary footwear. Not so the marshes. These invariably occur where the high water table is persistent; to collect the attractive wild flowers of these areas will almost certainly call for bare feet, old shoes, or boots, plus a lack of sensitiveness to mud. Marsh plants, if their home site is drained, soon die in soil that still has ample moisture for upland plants.

Almost everywhere the marsh vegetation passes over into the aquatic plant life of the deeper water (Fig. 18). A convenient distinction between the two is that of growth above the water's surface. If the leaf-bearing stems or the leaves extend above the water, we regard such plants as belonging to the marsh community. Those whose leaves are floating or submersed will constitute the aquatic community which is treated in Chapter IV. This distinction will, in the field, hold true to a very successful degree for the shoreward, shallow water or true marsh plants as distinguished from the pondward, deeper water aquatics.

FIG. 18. THE AQUATIC SOCIETY HERE IS BEING GRADUALLY ELIMINATED BY THE INVADING MARSH PLANTS. PICKEREL WEED IN THE FOREGROUND.

FIG. 19. WHERE LIZARD TAILS HOLD THEIR ANNUAL DANCE IN THE FOREST SHADE.

Every fresh water marsh will show more less distinctly the phenomenon of zonation. Certain plants prefer the shallower borders while others are best adapted to the deeper water. In other situations the character of the bottom soil exerts an influence.

Among the principal plants or dominants in these various marsh communities are the cattails *(Typha latifolia* and *T. angustifolia),* the arrowheads *(Sagittaria spp.),* the pickerel weed *(Pontederia cordata),* the wild rice grass *(Zizania aquatica),* its near relative the marsh millet *(Zizaniopsis miliacea),* the parrot's feather *(Myriophyllum sp.),* the lizard tail *(Saururus cernuus,* fig. 19), and many others. Scattered in these communities are relatively large numbers of attractive wild flower plants, which, though they are not dominant, are sufficiently common to add distinctive notes of beauty to the marsh-scape.

Among the more notable of these species which occur in a scattered fashion are: the marsh irises *(I. prismatica, Caroliniana* and *versicolor),* false pimpernel, which forms low local masses of vegetation along marsh borders, the exquisite marsh blue-bell, of the eastern marshes, a species *(Viorna crispa)* of clematis, which dangles its simple blue flowers over the neighboring plants, the marsh day flower *(Commelina hirtella)* of a deep sky-blue color, and the stately marsh skull-cap *(Scutellaria lateriflora).*

Both in water and on mud the peculiar golden-club spreads its long, simple leaves and holds its yellow club erect on a white stalk. The related primrose willow *(Jussiaea)* and false loosestrife *(Ludwigia)* are both showy in their floral offering. A number of marsh buttercups are found, differing more in leaf characters than in the flowers. Especially to be mentioned is the stately dragon-head *(Physostegia)* with its compact, very attractive spike of bilateral flowers, and the other mint which is called herein the "beautiful mint" *(Macbridea pulchra)* with its large rose-colored corolla tinged with purple and white streaks.

In wet or moist soil and to be included here are such species as the atamasco lily (*Zephyranthes,* fig. 143), which is not a true lily but just as beautiful as any species of the latter genus. Occasionally one finds the spider lily *(Hymenocallis)* with its exquisite veil-like membrane connecting the stamens. The green arrow-arum (Fig. 129) is a coarse-leaved herb, the leaf resembling that of the arrowhead but its inflorescence is partly enclosed by an enveloping spathe.

Of especial interest near the sounds is the "intelligence plant," a rather insignificant little herb with a leaf shaped like a thumb-nail and about twice as large. The sages of India are reported to teach that this cosmopolitan plant has the remarkable property of stimulating the intellect—that tea made from the dry leaves will make morons into masters and, it is to be supposed, college failures into Phi Beta Kappas. One person, to the author's knowledge, came from Los Angeles to the shores of North Carolina to restock his supply of these leaves so as to preserve a high mental tone. Outside of discovering that the leaf has a very queer taste the author can make no statement concerning this strange oriental idea. He can only wish it were true, being a teacher by profession.

Marsh plants exhibit some interesting characteristics which anyone may observe. Upon cutting into the stems and certain leaves as well, the student will

FIG. 20. CATTAILS BY THE MILE IN THE FRESH WATER OF CURRITUCK SOUND. VIEW FROM WHALE'S HEAD LIGHTHOUSE.

find the structure to be porous. Under the microscope these plants appear as if they were honeycombed with more or less regular passages. If one could become a tiny mite, like a red-bug, he could make a tour of these green lighted caverns above the water, later reaching the darker corridors below the water and the pitch black ones below the soil. Such an excursion would end in the smaller roots, which are being kept alive by oxygen brought through these passages by the gas diffusion process. Furthermore, carbon dioxide is being eliminated at the same time, for this intricate system of cell interspaces always connects with the outside air through the leaf and stem pores (stomata) which are to be found by millions in the plant parts above the water's surface. It is of interest to add that in many marsh plants these pores are totally incapable of being closed, a function readily occurring on the leaves of land plants.

Some of the largest of our marshes are found on the borders of fresh-water sounds such as the Currituck Sound in the northeastern corner of the state (Fig. 20). The great cattail marsh of that body of water near Whales Head Light is especially notable.

There is little development of marsh vegetation in the piedmont and mountains for the simple reason that in those areas all ponds and lakes are artificial ones and therefore small in size. These regions have been carved by erosion, and, in such areas, nature takes care never to leave any basins; everywhere the water runs down hill and away. As will be pointed out in the chapter on community relations, the coastal plain presents the interesting situation of the swamp forest competing for the marsh areas. From that discussion it will be noted that inland the conditions tend to favor the swamp forest so that most shallow water areas carry the tall woody vegetations rather than the low marsh type.

The phenomenon of marsh succession may in many places be observed in the lifetime of an individual. Due to filling in, with resultant changing of water depth, the vegetation zones will shift until the wet soil sedges hold forth where once the deeper water arrowheads grew. The movement is almost invariably that of a progressive shallowing of the water. On a sinking coast line the reverse changes would occur.

The habitat for the typical marsh plants is a triple one. The upper part of the plant is surrounded by air, the lower portion of stem or leaves is in the water, while the underground stem and roots are embedded in water-logged soil. Many plants in their growth develop strikingly different leaves in the water as contrasted with those out of it. The aerial leaves invariably have broader blades or blade segments, while the submersed ones tend to be much dissected into fine or slender lobes. The roots of many of these plants do not produce root hairs, a deficiency probably to be correlated with the perennial abundance of water.

CHAPTER III

THE "WET" TREES, OR SWAMP FOREST

In our swamp forests we have, as in the case of the sandhills, a plant community which is distinctly southern. In the shallow waters of ponds and river borders of the north only shrubs and low willows are to be found. On low moist grounds which are occasionally flooded for brief periods, such tall trees as the ash, elm, and maple may occur, but such areas are not comparable to the southern swamps, where in years of ordinary rainfall the water stands a few inches to two feet deep much of the time.

Three genera of trees, the gums, the cypresses, and the white cedars are the dominants of our swamps. The relative abundance of these in any area gives the character of that particular forest. Very frequently any one of the three may be seen in nearly pure stand so that the locality may be definitely designated as a gum, cypress, or white cedar swamp.

The Gums

The swamp gum *(Nyssa biflora)* and the tupelo gum *(N. aquatica)* are the only trees of this genus adapted to the water-covered soil. They differ somewhat in choice of habitat, the tupelo gum tolerating the deepest water while the swamp gum ranges into the shallower water and may occasionally be seen in the

FIG. 21. THE "WETS" OF THE TREE WORLD. SWAMP GUMS ARE THRIVING HERE IN 18 INCHES OF WATER.

shrub bogs, even surviving as a relict in grass-sedge bogs. The gum forests are very abundant along the margins of the coastal plain rivers. Because of the broad leaves produced luxuriantly in their crowns these forests are very dark below, a condition tending to hold in check the minor vegetation which might otherwise enter. Both of the gums prominently exhibit the character of swollen trunk bases (Fig. 21), a character which arouses one's curiosity as to its significance.

It is very easy to observe on a cut base that the wood ring number is the same as that above the base. The enlargement is thus due to the increased thickness of the rings which is a response to the proximity of

FIG. 22. THE CYPRESS PATRIARCH OF WHITE LAKE, WITH HIS LUXURIANT BEARD OF SPANISH MOSS.

the water. Since the soil and subsoil of most swamps are of a soft silt or muck nature, it is readily apparent that these wide flaring bases with their still wider spreading roots give the trees a degree of anchorage which they would not otherwise have. In the larger trees the bases are hollow.

THE CYPRESSES

The cypress *(Taxodium)* has been divided into two species, though the differences between them are not great. *T. adscendens* has the tiny leaves laid forward along the slender dwarf branches while in *T. distichum* they extend outward from the delicate twig. The trees thus have a different aspect, which may be noted at some distance. It is also believed the species differ in habitat, the first mentioned preferring sandy soil or

FIG. 23. CYPRESS KNEES SHOULD BE CALLED CYPRESS LUNGS INSTEAD, FOR THROUGH THEM THE GAS INTERCHANGE NECESSARY FOR THE LIFE OF THE ROOTS, IS CARRIED OUT.

subsoil, while the latter is more frequently found growing in silt or muck bottoms. When the cypresses are seen in virgin stand (of which few are now left) they make a most impressive picture. Like the gums, they have swollen bases (Fig. 22) which in the old and large trees of *T. distichum* become prominently buttressed as well. The trunks or boles are long and slender, carrying at the top the great feathery crown. The larger trees reach a trunk diameter of fifteen feet, and an occasional one is larger than that.

To the newcomer in a cypress swamp, the woody blunt-ended upgrowths from the roots are so unusual that the visitor will immediately ask about them (Fig. 23). It would be much more accurate to inform such an inquirer that they are lungs rather than "knees," the name by which they are usually known. Investigation has shown that a vessel placed over the top of the spongy-barked knee will rapidly increase in its carbon-dioxide content and at the same time lose oxygen. Like a true lung this structure is an organ where gas interchange takes place, a function unquestionably of significance in relation to the life of the roots which are buried in a stratum characterized by a very low oxygen content.

Observation and experiment have shown that cypress may become established only in swamp areas during drouth years when there is no surface water present. Without a high oxygen supply the seeds cannot germinate. In the cypress life story we have an excellent example of the biological rule that adult organisms can survive under conditions which are death-dealing in the early stages of development.

The White Cedar

The white cedar (*Chamaecyparis thyoides,* fig. 24) is an interesting tree with its spray-like twigs resembling the arbor-vitae while the bark is suggestive of the common red cedar. So valuable is its wood that vast acreages of it have been cut over. So thoroughly has it been removed that it can no longer be said to be a common tree. As to habitat, it avoids the deeper water; in fact, much of it is found associated with yellow poplar, an upland forest tree, which in the east occurred on very shallow swamp lands. According to Dr. C. F. Korstian of Duke University, the densest original stands were upon peat where drainage did not allow the accumulation of any considerable depth of surface water. It is of interest to note in this connec-

FIG. 24. THE "COME-BACK" OF WHITE CEDAR IN THE DISMAL SWAMP. NOTE EVEN-AGED CHARACTER OF THE STAND.

tion that the peat bog known as the Open Grounds in Carteret County, contains near the base of the peat, innumerable logs of an ancient white cedar forest.

Many minor trees are characteristic of our swamps. Among these are the swamp red maple *(Acer rubrum* var. *Carolinianum)*, the pumpkin and pop ashes *(Fraxinus profunda* and *F. Caroliniana)*, the swamp hickory *(Hicoria aquatica)*, and the planer tree *(Planera aquatica)*. On the borders the sweet gum *(Liquidambar styraciflua)* is very common.

ON THE SWAMP MARGINS

On the swamp margins, two vines which climb high in the gum and other trees, are especially noteworthy. These are the supple-jack or rattan *(Berchemia scan-*

FIG. 25. THE CAUSE OF SUCH GOUTY BRANCHES ON A TREE IS OFTEN PUZZLING. THESE WERE CAUSED BY MISTLETOE ON SWAMP GUM.

dens) with simple ovate, stalked leaf-blades borne on green stems. The more common cross-vine *(Bignonia capreolata)* with its four-leaved clusters hanging from the nodes, is unusually interesting on account of the plus sign (+) recorded in the stem structure, which is made evident when the stem is cut across.

Many shrub bog plants enter the shallower swamps, where they constitute an under-story of vegetation. Of these the most common is the ti-ti or leatherwood which is most tolerant of water-logged soils. On the swamp borders any or all of the other bog shrubs may be expected. Where the land slopes sharply into the swamp water, the bog shrub zone may be but a few feet in diameter but it will be very definite.

In addition to these shrubs, a few frequent the swamp borders or shallower swamps which are not

FIG. 26. WHERE THE ROAD PLAYED FIREMAN IN THE MIDST OF A SWAMP FOREST.

commonly seen in the bogs. One, which everybody should know so as to avoid it, is the poison sumac *(Rhus vernix)*. It has long pinnate leaves with leaf stalks generally red in color. The swamp rose *(Rosa Carolina)* is a superb wild rose and always a picture wherever found in bloom. One of the dogwoods *(Svida stricta)* is very frequent, but more commonly seen are the two or three species of *Viburnum* of which perhaps the possum haw *(V. nudum)* is most common.

Practically no herbaceous flora is to be found in the swamp proper. In the drier areas the chain fern *(Anchistia Virginica)* may be expected, a fern which somewhat resembles the cinnamon fern. Because of the almost complete absence of herbaceous wild flowers from the swamp forest, no key is included for that community in the key sections.

The habitat of the swamp forest is in part similar to that of the bogs, except that it is wetter both as to amount of water and as to length of the period of flooding. Only in the occasional drouth year do the swamps lose their surface water. Even then the soil ordinarily remains wet enough to keep the vegetation alive. The state of excessive water of the swamps is of course related to the topographic factor or "the lay of the land." Swamps are uniformly lowlands representing areas into which the surrounding country drains. Bogs, on the other hand, are uniformly poorly drained uplands. Only the lowlands can accumulate and hold surface water; the uplands can never do this. Hence in topography we have the key to the wetlands of the coastal plain.

In such wet places as swamps, about the last thing to be expected is a fire. Yet during drouth periods some of the very worst conflagrations are those which sweep through the crowns of cypress and gum trees (Fig. 26). Millions of dollars worth of desirable timber have been lost through swamp fires alone.

CHAPTER IV

FISH GARDENS: AQUATIC VEGETATION

Everyone, of course, is familiar with the green scum which appears on ponds and pools, and almost everyone knows that such masses are made up of microscopic plants very low in the order of nature, whose ancestors have always lived in the water. It is not so well recognized, however, that the flowering plants with which we are concerned and which have made themselves at home with the fish were, once upon a time, land forms. Like the seals and whales which descended from the land mammals, our aquatic higher plants present a long history of progressive adaptation to submersion until, at the present time, many of them are able to grow and reproduce by seeds, though they are entirely under the water all of the time.

Fresh-Water Plants

We shall here be preëminently concerned with the fresh-water forms, which far exceed the marine submersed plants in number and interest.

Like the marsh plants, the aquatics find their greatest development in lakes, ponds, and fresh-water sounds, and estuaries of the eastern part of the state. And let it be stated at once that in the eastern area, the shallow fresh-water bodies are extraordinarily rich in the water-loving vegetation. When it is realized that ducks and geese are dependent for their livelihood upon aquatic plant life and marsh plants, the abundance of this food is reflected in the fact that the sound and lake regions of the low country of North Carolina constitute some of the best hunting grounds in America.

To study these interesting green flowering plants which have changed their homes from air to water, one must either have a boat or a bathing suit. Since many of these plants are rooted six to ten feet under the surface, to obtain the whole specimen demands a little diving ability.

Due to the difficulties involved in observations upon them, a great deal yet remains to be learned about their distribution. We know, however, that like the marsh species they respond to differences of water depth and of soil in the case of those which are rooted. To illustrate the latter point, certain pipeworts and spike rushes have been found to occupy the sandy bottoms, while the water weeds, water lilies, and many pondweeds distinctly avoid the sand, much preferring the dark silty or muddy substrata.

In this community, as it has already been pointed out, we may arbitrarily include those plants which do not hold their leaves or grow the leaf-bearing stems out of the water. These are the extreme water lovers among plants. In the case of the totally immersed types it goes without saying that everything they obtain comes from the water and the water-soaked soil below. The oxygen for respiration, the carbon dioxide for food making, the plant nutrients so necessary for all functions in metabolism, all are in solution in the water and the plants are able to take them in directly at all points since these submersed forms are devoid of that thin waxy layer which covers the green leaves and young stems of land plants.

The Effects of Submersion

Another interesting character of aquatic plants is brought out when, in imagination, we submerse land shrubbery, trees, and herbs, and then imagine what the waves and water currents would do to them. Instead of meeting the water habitat by developing rigidity, the aquatics have done just the opposite and, paradoxically, have conquered the heavier medium through weakness. Lift any water plant into the air and it will be found totally incapable of holding its head up, so to speak. These plants merely developed enough buoyancy so that they could keep near the surface, but to the water movement and waves, the supple stems offer no lateral resistance. The under-water leaves are typically slender and finely divided, so that when the plant is tending to be washed away by a strong current, there is little surface exposed on which the water may get a grip.

Many species with broad floating leaves bearing leaf pores on the upper side, will have different submersed ones which in no instance ever bear leaf pores. The common, arrow-leaved spatterdock *(Nymphaea sagittaefolia,* fig. 27), of the black water coastal plain rivers and ponds, is a plant of this type. The contrast between its shapely, floating leaf and the formless lettuce-like leaf below the surface is surely a great one. It is difficult to believe the two grow from the same stem buried in the bottom mud.

FIG. 27. THE ARROW-LEAVED SPATTER DOCK GRACES THE STREAM AND POND BORDERS OF THE COASTAL PLAIN. NOTE SUBMERGED LEAVES OF THE SAME PLANT WHICH ARE VERY DIFFERENT FROM THE FLOATING ONES.

THE BEST OF ALL THE DUCK-FOODS

Deserving especial mention is the best of all the duck-food plants, an aquatic which because of its abundance in certain North Carolina waters has made of these a hunter's paradise. This plant is one of the pondweeds *(Potamogeton pectinatus)*. The leaves of this species are all of the very slender or filiform, submerged type. The small flowers are borne just above the surface but the fruit matures below. This fruit, a little plum-like structure nearly three-eighths of an inch long, together with the leaves and tuberous roots, is especially attractive to the game birds which in enormous numbers spend their winters in the sound country.

The Smallest Flowering Plant Known

Also to be added here are a few species which are of interest in themselves. The smallest flowering plant known, the dwarf duckweed *(Wolffia Columbiana)* is a tiny mass of green tissue about the size of a pin-head which drifts about on the surface of quiet waters. The remarkable tape-grass *(Vallisneria)* bears the seed-bearing flower on a long, slender stalk which reaches to the surface. The staminate flowers are produced at the bottom of the water and when mature are cut loose, whereupon they float to the top, open the anthers, and drift about by chance into the carpellate flowers where the pollen is by contact deposited upon the protruding stigmas. After pollination the stem shortens by contracting into a spiral and pulls the flower beneath the surface for the period of fruit development. Very common everywhere in our eastern fresh waters is the bladderwort *(Utricularia)* represented by many species. These all have little traps which operate upon the trap-door principle. Tiny water-forms of animal life are enticed within them, when down comes the trap-door, jailing them most effectually, and then the digestive juices appear within, to finish the process most advantageously for the bladderwort plant.

Characteristics of the Habitat

The habitat of the aquatic plants is essentially similar to that of the marsh as far as the submerged and rooted portions are concerned. Only the upper side of the floating leaves which bears the leaf pores, is adapted to the aerial habitat. The low oxygen content of the submerged soil is reflected in the air-bearing spongy tissues of many of these plants. The ready interchange of gases and mineral salts dissolved in the water, has already been mentioned. Above all in importance in connection with the aquatic habitat is the factor of permanency. Even though rooted in damp soil, if the surface water for any reason is drained away, the plants will quickly die, so completely dependent have they become on the water environment.

A Distinctive Kind of Peat

It is a fact worth knowing that aquatic vegetation forms a distinctive kind of peat. The middle layer of peat in the Open Grounds of Carteret County is of this type, telling the story of a former impounded shallow water lake which once occupied the area now covered by a shrub bog. Visitors to this great pocosin lying twelve miles north of Beaufort, may recognize this middle peat type by its soft, sticky nature, which makes it resemble black car grease more than anything else. In this region as elsewhere, the student of peats can easily read the story of changing water levels which near the coast means changing land levels in relation to the sea. The minor recent ups and downs of the land there are all perfectly recorded through the changing vegetation producing different kinds of peat.

CHAPTER V

WHERE WINTER NEVER COMES: THE EVERGREEN SHRUB BOG

THE WRITER, in company with a companion and a guide, once walked for six hours through black mush. It was early April when most of the trees and shrubs were clothing their branches with leaves; yet on the "mush" which we traversed was a thick cover of evergreen, broad-leaved bushes from knee to waist high (Fig. 28). One could at the frequent stops look across this rich green shrub cover, and had it not been for the sharpness of the air, the month might as well have been July. We were going into one of the great evergreen shrub-bogs of eastern North Carolina (Angola Bay in Pender County) to visit the virgin long leaf pine trees which stand on a small sand island in the central part of the area. But long before we got to the pine trees, we found this wide, monotonous morass a thing impressive and distinctive in itself. The deeper we got into it both horizontally and vertically (for in spots we went thigh deep) the more of a "lost world" aspect it attained in our minds. Had one of those Mesozoic reptiles approached through the thin stand of pocosin pines, floundering his uncanny body amid this peatty mire, we should not have been greatly surprised, so perfectly would such a beast have fitted this unique habitat.

FIG. 28. AWAY TO THE HORIZON EXTENDS THE SEEMINGLY ENDLESS SOCIETY OF BOG SHRUBS WITH ITS OPEN STAND OF POCOSIN PINES.

The early explorers and settlers in the flat lands of the lower three coastal plain terraces were much impressed by this broad-leaved shrub vegetation which did not shed its leaves in winter. Scattered in smaller and larger masses, such a plant community was readily singled out, especially in the cold season, as a special sort of growth, and became known by two names. These names not only referred to the plants but included as well the wet soil habitat which, as will be fully explained later, has peculiarities and variations of its own. Both of these names, "pocosin" and "bay," have come down to the present and are part and parcel of the verbal currency in eastern North Carolina.

An early description (1737) of these areas by John Brickell is worthy of quotation, even though he has not

FIG. 29. CANE, CANE AND YET MORE CANE, A RISKY PLACE IN CASE OF FIRE. ANGOLA BAY, PENDER COUNTY.

clearly distinguished swamp from pocosin. Note the old spelling of pocosin. "Other Lands in this Province are Perkosons, where large Cypress Trees grow, others Swamps, where hollow Canes, Myrtle Trees and several sorts of Vines grow, and produce good Pasturage for Cattle, but generally the Habitation for wild Beasts; both of these being very wet and low lands and so full of Cane and Underwood, that there is no passing through them, many of which are several miles in length. The Indians in their Hunting Matches set these Places on fire at certain seasons of the Year, by which Means they drive out the Games and kill vast numbers of them."

Upon seeing the evergreen pocosins of southern Virginia and the Carolinas, the northern winter tourist exclaims "The Southland." He recognizes immediately the summer-like aspect of this massed greenery and even though the temperature may be well below freezing, the sight of this landscape loaded with verdure will do much to make his feelings belie the thermometer. Our "baylands" or "bays" present the sharpest contrast with the more northerly woodlands, where only the occasional pines add a touch of color to the monotonous gray of the uplands. But how different the southern pocosins with their deep, solid, rich greens! And yet the frost has hit them many times before midwinter is reached.

To be included in the shrub bogs are the great areas covered by the grass shrub commonly known as cane or reed. Large portions of Angola and Holly Shelter Bays are in solid cane (Fig. 29). However, a more detailed treatment of the canebrakes will be taken up later when the shrubs are discussed separately.

The Homeland of the Bog Shrub Garden

The soil habitat of the bay or shrub bog is generally a flat upland which is water-logged for long periods when the rains are frequent but may become fairly dry during the latter part of extended drouths. Ditches and streams always carry free water away from the bogs into the lower areas, where the water accumulates on the surface, making a marsh or a swamp, which latter will carry gum and cypress trees. (See chapter on the Swamp Forest.) Thus the shrub bog site is one that is poorly drained but never carries water of measurable depth on the surface. The poor soil drainage is due chiefly to three factors: the flat topography preventing a rapid run-off, a fine soil texture preventing rapid lateral movement underground, and a non-draining subsoil preventing downward drainage, all of which contribute to developing a high water table in seasons of moderate rains. If the area between streams is very broad (five to twenty miles) the soil may be composed of medium to coarse sand; yet poor drainage will result just the same.

If the high water table were permanent the marsh plants like cattails and sedges would have a chance to compete for such areas. The fluctuating character of the water table, however, easily accounts for the continual elimination of these plants. Even so, occasionally one sees a lonely arrowhead or cattail temporarily established in a low spot of the shrub bog.

The nature of the soil as to whether it is mineral or peat (nearly pure organic matter) is of no significance; if it is wet for weeks at a time, then medium dry during the rainless times when the water table falls many feet below the surface and too frequent fire has not destroyed the vegetation, such soil will carry the evergreen shrub complex. Commonly bogs are thought to be peatty in composition, but this cannot be true when the same plant society on our southern peats lives just as well when rooted in wet sand. Thus under southern conditions a bog may accurately be defined as a flat upland with a fluctuating water table, this latter factor, when high, being at or near the surface.

One interesting exception must be noted here in passing. In the high sandhill district of Harnett, Moore, Cumberland, and Richmond counties, buried clay-like, impermeable layers, cause the water in many places to flow out on the hillsides, so that for long distances the lower slopes are kept perennially wet by seepage. On these slopes which are often very steep,

FIG. 30. THE OPEN GROUNDS, ONE OF NORTH CAROLINA'S GREAT SHRUB-COVERED PEAT BOGS. THE BUILDINGS ARE THOSE OF AN AGRICULTURAL DEVELOPMENT COMPANY.

shrub-bog vegetation is found with most of the characteristic species represented. Such hillside bogs are very unique.

Any discussion of the bog habitat must contain a word about the low oxygen content of bog soils during the flooded period, because in this condition is found a possible explanation why other plants cannot make themselves at home in bog areas. Plant roots may drown as easily as an air breathing animal; the root cells must have oxygen. For ordinary broad-leaved shrubs like those of our yards, the root system must be wide ranging and reasonably deep if sufficient water is taken up to keep pace with the water loss from the leaves. It is at once recognized that if there is something which interferes with root growth and maintenance, the amount of water entering the plant will be cut down and unless the plant meets this deficiency by checking water loss above ground, its life will be endangered. This "something" in bogs is low oxygen during the long soggy soil periods when water displaces the air in the pore spaces. The restricted evaporation of water from aerial parts of bog plants is a well known phenomenon to botanists. And while careful measurements of the water outgo in the case of our bog shrubs have not yet been made, there can be little doubt that the thick waxy coatings on their leaves and other characteristics which need not be enumerated here, all conspire to check water loss so that the poorly developed root-system, in its low oxygen environment, may at all times keep the all-important water content from falling to the danger point.

The water in the surface soil layer is not only tending to drown the roots but it succeeds to a high degree in drowning billions of those invisible denizens of soil, the bacteria and fungi. This situation results in the slowing up of the decay process, for, as everyone knows, when you keep micro-organisms away from organic matter, there is no decay, as the vast supplies of canned goods testify. Hence it is that bogs are, in a sense, one of nature's canning factories, where plant products are preserved. These remains accumulate in the dark colored layers which carry the latest vegetational growth on top, and so a process of building up peat goes on. In Angola Bay the organic remains reach a maximum thickness of nine feet. In the Dismal Swamp the organic deposit reaches twice this figure.

Not only the bacteria causing decay are checked in

FIG. 31. AN ENTRANCE TO THE GREAT HOLLY SHELTER BAY, ONE OF THE LARGEST OF THE SHRUB BOGS. NOTE POCOSIN PINES.

bog soils but the valuable nitrate-forming kinds are also completely held in abeyance. Every gardener and agriculturist is familiar with these most important helpers which are able to turn the ammonia in manure to nitrates, a fertilizer which constitutes a most favorable form of nitrogen. It is interesting to know that these bog soils have no nitrates produced in them and the plants growing there are absorbing their nitrogen in the guise of ammonia compounds. The ammonia, as in the case of manure, comes from the slow breaking down of the protein portions of the dead plant debris.

A note on the distribution and occurrence of shrub bogs will not be amiss. They are very abundant on the Chowan and Pamlico terraces, the two lowest near the sea. Among the larger of these upland poorly drained areas are large portions of the Great Dismal Swamp. The following counties have extensive areas of this type of land: Brunswick, Carteret, Columbus, Dare, Hyde, Onslow, Pender, and Tyrrell.

As would be expected, the borders or transition zones around the eastern swamps and marshes will carry more or less of the bog shrubs. In such localities there will appear a telescoping of the swamp forest trees and the bog shrubs. This overlapping of two communities need not be confusing, chiefly because our shrubs for the most part remain shrubs and need never be mistaken for the tall trees of the swamp. One must look out particularly for the swamp gum *(Nyssa biflora),* which prefers much the same habitat as titi and when it is small in size may prove puzzling since we have not included it in our list of shrubs. A description of it may, however, be found in the chapter on the Swamp Forest.

The effects of fire have already been mentioned. A few additional observations must be included since they are frequently seen by anyone interested in bog plant life. In the case of peat bogs, if fire comes when the peat is wet and the lower parts of the shrubs remain uninjured, the plants may successfully regenerate and the shrub cover holds on, but it is not so tall as formerly. If, on the other hand, the drouth has been long and the surface peat is dry it will burn and with it will go the plants, leaving areas of variable size and shape, bare blackened wastes, colored here and there by sickly masses of gray where the rain has washed the ashes into pockets. On such places in summer there will come splotches of finest textured green due to the almost magical growth of the "fire moss" *(Funaria hy-*

FIG. 32. WHERE THE ROAD CHECKED THE FIRES; SHRUB BOG ON LEFT, GRASS-SEDGE BOG ON RIGHT.

grometrica). But a much larger and more permanent succession is that of the coarse golden rod *(Solidago fistulosa).* This is sufficiently common after fires in peat lands to deserve the name of "fireweed."

Where the shrub cover has been on mineral soil but removed by frequent fire, the grass-sedge bog society or savannah will take its place (Fig. 32). This does not occur on peat, apparently for the reason that fire destruction of the soil itself readily eliminates the shallow roots of the herbaceous plants. Hence one seldom sees typical grass-sedge bog on peat. For a fuller account of these relations the reader may refer to the chapter on the Savannahs.

It should be repeated that this unique evergreen plant community, which grades on one side into the deciduous swamp forest and on the other into the deciduous drained upland forest, is one of the most interesting natural plant aggregations in North Carolina. Not only are most of the species evergreen but they possess the simple, pointed type of leaf which is so characteristic of tropical jungle trees (Fig. 33). Some ten of them are heaths *(Ericaceae)* being related to the familiar blueberries. All are remarkable in their ability to tolerate water-logged soil for periods of many weeks, though, as already pointed out in specific cases, the species vary somewhat in this regard. The plant lover will find many a pleasant hour in the company of these shrubs, a number of which he can dig up and take home with him, where they will prove desirable yard acquaintances the year round.

Bog or Pocosin Shrub Portraits

Returning to the remarkable shrubs which make up this unique plant society, we shall discuss them separately, thus making it possible for the reader to identify these exceptionally interesting plants. Since certain of these shrubs may be successfully used in landscape work, detailed information concerning them will not be amiss. Particularly is this statement of significance when it is realized that they are hardly ever seen in plantings, even though they are available in eastern North Carolina in unlimited quantities. A New York nurseryman who specializes in shrubs once sent to England for the honey-cup *(Zenobia)*, where it had been taken from the Carolinas many years ago. He had been totally unable to make contact with anyone who could obtain the plant for him from its native heath. And yet on many of our large bogs it is one of

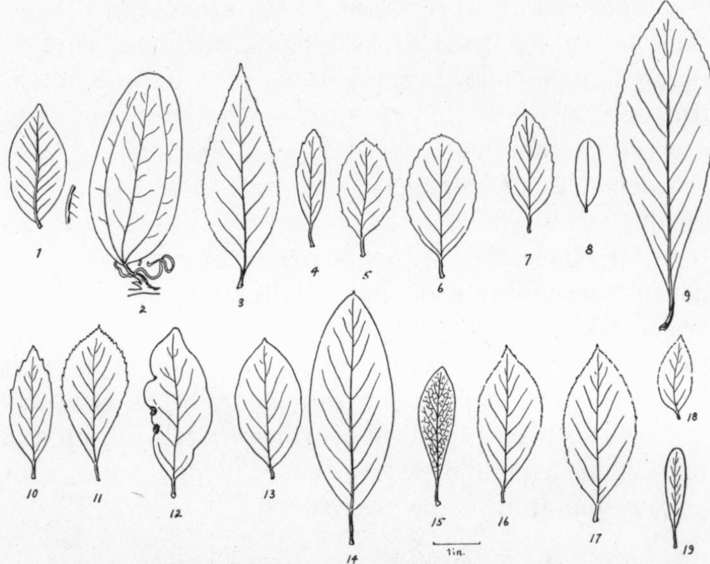

FIG. 33. LEAVES OF TYPICAL POCOSIN SHRUBS: 1. Angle-stemmed fetter bush or hurrah bush (Pieris nitida). 2. Bamboo-brier (Smilax laurifolia). 3. Bog dog laurel (Leucothoe axillaris). 4. Gallberry (Ilex glabra). 5. Shining or swamp gallberry (Ilex lucida). 6. Honey-cup (Zenobia cassinifolia). Blunt leaved variety. 7. Honey cup, pointed leaved variety. 8. Leather leaf (Chamaedaphne calyculata). 9. Loblolly bay (Gordonia lasianthus). 10. Wax-myrtle (Myrica Carolinense). 11. Pepper-bush (Clethra alnifolia). 12. Red bay (Persea Borbonia). 13. Stagger bush (Pieris Mariana). 14. Sweet bay (Magnolia Virginiana). 15. Titi (Cyrilla racemiflora). 16. White ozier (Leucothoe elongata). 17. White ozier (Leucothoe racemosa). 18. Whitewood (Xolisma foliosiflora). 19. Wicky (Kalmia cuneata).

the commonest species. From the Open Grounds in Carteret County many train loads could be shipped were the demand great enough.

It must be recognized at once that these bog plants cannot be grown in the drier places in one's grounds. They must be given the moister low areas and if possible they must enjoy the protection of partial shade. Comments along this line will be made in connection with each species.

For convenience in reference the shrubs will be arranged alphabetically by common names.

ANGLED-STEM FETTER BUSH *(Pieris nitida)*. The shrub is easily identified by its leathery leaves which have curiously raised midribs (Fig. 33); instead of a groove marking the midrib on the upper side, this structure actually projects above the leaf surface. In addition the twigs are sharply angled and commonly are of a reddish hue. The flowers are borne in close clusters from the leaf axils and are tinted light red at first.

This shrub is well worthy of trial in the eastern half of the state. It should be given as moist a place as possible with some shade.

BAMBOO BRIER *(Smilax laurifolia)*. No plant is more characteristic of the shrub bog than this coarse species of green-brier. It is the nemesis of deer hunters from whom it has extracted untold numbers of unprintable expletives. With its tough, viciously thorny long stems lying across every trail it becomes a highly successful impediment to progress. Only a heavy bush knife can conquer it and this important item of equipment should be included on any expedition into the great pocosins of the state. Like its associates it is evergreen. It may be distinguished by its large elongated-elliptical leaf (Fig. 33).

BOG DOG-LAUREL *(Leucothoe axillaris)*. In the mountains the dog hobble *(L. Catesbaei)* is an attractive low shrub with showy compact clusters of white flowers in the leaf axils. The eastern twin of this mountain species is to be found in the dog-laurel which has the taper-pointed leaf (Fig. 33), though the leaf is larger, and the same many-flowered racemes partly

FIG. 34. THE FLOWER CLUSTERS OF THE BOG DOG LAUREL HIDE UNDER THE LEAVES.

hidden beneath the spreading leaves of the arched branches (Fig. 34). This shrub may reach a maximum height of seven feet but is almost invariably smaller because of fire.

Like many other heaths *(Ericaceae)* it is poisonous to stock. Pastures should never make contact with shrub bogs because of the numerous species of these plants which may cause serious losses to the stock grower.

This is one of the best of the bog shrubs for transplanting. If the soil does not become excessively dry it should do well in grounds throughout the coastal plain and most of the piedmont.

CANE OR REED *(Arundinaria tecta).* Two species of cane have been given in the books based upon the way the flowers were borne—from the underground and aerial stems respectively. It has recently been shown that the same plants can produce the flowers both ways; so we may be sure that only the one species exists. As already indicated, cane is a grass shrub and, true to the bog type, is evergreen. If fire does not set it back it may grow to twenty-five feet.

An especially interesting fact, not generally known, is that the life of cane is determined in its heredity. After a certain number of years (about fifteen) the plants flower and die, root, stem, and branch. If the cane is all from the same parentage large masses of it will disappear as if by magic. The myriads of seeds left behind, however, soon sprout and in a few years the brake is much as it was. This same law of determinate growth is also true of the tree-like bamboos of the Orient, which are, incidentally, near grass relatives of our cane.

Where cane becomes established (Fig. 29) it holds its own against the other bog shrubs and may even encroach upon these through root competition. Its subterranean stems become so massed as to make this plant the leading humus or peat builder of the southern bogs. In traversing these areas one quickly notices that when he enters the cane the walking becomes much drier, for the matted interlacing stems build themselves slightly above the water table.

No ground fire in the coastal plain district, a region noted for its fires, can equal that of a tall canebrake. Under a high wind the flames leap forty feet into the air and the cane stems crackle as loud as musketry. It is a most spectacular sight but one seldom seen in the big canebrakes since no one dares risk close proximity

FIG. 35. THE GALLBERRY "HONEY FLOW" IS ON.

because of the danger of becoming enveloped by a circle of fire.

CHAIN FERN *(Anchistea Virginica).* So omnipresent is this tall fern scattered over the shrub bogs that mention of it must be made. If one knows the more familiar cinnamon fern of moist spots, he will be apt to confuse the two. A glance at the underside of the older leaves of the bog fern will disclose the fruiting spots arranged along the midrib of the leaflets in chains. In the cinnamon fern the fruiting structures are borne on a special reduced set of leaf-like structures produced in the midst of the circle of true leaves. Closer observation will disclose a characteristic venation of the chain fern.

CHOKE BERRY *(Aronia arbutifolia).* This little relative of our apple trees is widely scattered in the bogs and thus a very unimportant member of the community. The leaf margin is finely toothed and the underside is white with a soft close pubescence very distinctive for the bog shrubs, which have mostly very smooth leaves. In the spring it bears a few clusters of small apple-like flowers.

GALL BERRY *(Ilex glabra)*. This close relative of the Christmas holly is perhaps the best known of the shrub bog plants (Fig. 35). It is well named, for its black berry seems to be loaded with quinine. It may be recognized by its narrow wedge-shaped leaf with the few teeth it has being borne near the tip (Fig. 33). In the open the leaves extend outward and upward from all sides of the stem, while in the shade they appear to be flaring out from two sides. This latter aspect is an orientation to the weakened light, but it changes the aspect of the plant so much as to make it appear almost a different species.

Of especial interest is the fact that this is one of the most valuable honey plants in North Carolina. The nectar glands of its small white flowers are most productive, and when the "gall berry flow is on," as the apiarist speaks of it, the bees are "busy bees" indeed.

Because of its choice of the drier portions of the shrub-bogs, this low woody plant lends itself very successfully to cultivation. On north sides of buildings in soils that are not subject to extreme drouth, it should do especially well. During the winter when most of one's shrubs are in the leafless state, a mass of gall berry foliage would be highly appreciated.

HONEY-CUP *(Zenobia cassinifolia* and *Z. pulverulenta)*. In all of the bogs no more beautiful flowers are borne by any plant than the pendent bells of the honey-cup (Fig. 36). Developed profusely in long

FIG. 36. LIKE TINY-CHIMES THE HONEY-CUP FLOWERS SWAY IN SUMMER BREEZES.

FIG. 37. THE SHOWY LONG CLUSTERS OF THE HONEY-CUP SHRUB ARE ONE OF THE FEATURES OF THE POCOSIN LANDS.

terminal inflorescences, they make a most attractive picture, especially when one finds them in masses where the shrubs are thick. It is the blue-ribbon winner of our bog society (Fig. 37).

It gets its name from the story that the blossom has honey in the base, but due to the conformation of the flower, the bees cannot reach it. The reason given for tantalizing these seemingly harmless insects is that God, disliking the misbehavior of the bees in working on Sunday, made the attractive but useless honey-cup to punish them. It is to be suggested that there is as much doubt in regard to the first part of the story as the last. Like most of our bog heaths, this one is a very low producer of nectar.

Reference to the leaf diagrams will help in identifying this desirable form (Fig. 33). There are two forms of leaves as to shape, both of which are shown. The species *Z. pulverulenta* differs from the other one in having the underside of the leaf covered with a heavy white bloom which makes its appearance totally distinctive from all the rest. This latter plant is not common. It has been observed by the writer occurring in the small stream border bogs of the Sandhill district.

Both species are very desirable for landscape work if they can be given a reasonably moist place with partial shade. The author has grown this shrub at Raleigh on a north wooded slope for many years. It does not flower so profusely in such a place but the flowers one does obtain are so lovely as to make any effort to grow this species well worth while.

LEATHER LEAF *(Chamaedaphne calyculata)*. From the northern bogs, this little shrub has come to us with its distinctive leafy racemes of flowers. It seldom

FIG. 38. ONE OF THE MOST BEAUTIFUL SMALL TREES IN NORTH CAROLINA, THE LOBLOLLY BAY.

FIG. 39. THE FRAGRANT FLOWER OF LOBLOLLY BAY, A FLORAL PRIZE WINNER OF THE WILDERNESS.

FIG. 40. WHERE THE WAX MYRTLE HAS BECOME A TREE. NOTE THE VARIEGATED BARK OF THE TRUNKS OF THESE OLD "BUSHES."

grows more than two to three feet high and is not important from a gardening point of view (Fig. 33).

LOBLOLLY BAY *(Gordonia lasianthus).* Few of the woody plants in the upland wet areas attain tree size. This is one of them, but, of more importance, it is, perhaps, the most beautiful small tree in the state; its medium-large lustrous, evergreen leaves, together with its large white marvelously scented flowers, alone make it remarkable (Figs. 38, 39). In aspect it suggests a kind of magnolia, but it belongs to an entirely different family—the *Theaceae* or tea family.

Unfortunately, little success has been met in transplanting this most pleasing small tree. It is seldom seen in cultivation, even in the eastern counties where it is found. To enjoy it, one must watch for it along the roadsides in summer when the rich creamy-white

FIG. 41. PEPPERBUSH SPIRES ARE VERY FREQUENT ALONG BOG BORDERS.

flowers are scattered over the tree, like electric bulbs lighting up the deep green foliage.

PEPPERBUSH *(Clethra alnifolia).* Like the preceding this plant prefers the drier portion of the bog areas. In walking through the shrub complex one will note the clumps of pepperbush on the little knolls which extend just a few inches above the highest water table. The erect pointed racemes of the relatively large flowers, with their dark anthers giving them a peppery aspect, are very characteristic (Fig. 41). This low shrub is genuinely attractive when seen *en masse* with its numerous spire-like inflorescences pointed at all parts of the sky. In addition the flowers possess a very pleasing fragrance. It is an exception to the bog type in that the leaves (Fig. 33) are deciduous.

The pepperbush is readily transplantable and in

moist places should make a good growth without any shade whatsoever.

Pocosin Pine *(Pinus serotina)*. This tree, which because of fire is generally seen as a small tree, may always be expected wherever peat occurs. It will be found commonly in an open stand where the curiously crooked trunk shows plainly against the sky. The tree superficially resembles the loblolly pine, for which it is often mistaken, but it may be instantly recognized by the short, stocky top-shaped cones which hang on the limbs for many years after they are developed. This is the only pine which can tolerate the wet soil prevailing in the bogs for very long periods. It is remarkably resistant to fire, but even if the trunk is killed numerous shoots will, shrub-like, spring from the base, quickly carrying out regeneration. Because of its slow growth and commonly small size it is not regarded as a satisfactory timber tree.

Red Bay *(Tamala pubescens* and *T. Borbonia)*. This small evergreen tree is extremely abundant in most shrub bogs. The normal leaves have little to distinguish them, but since there is an aphid or plant louse which has the habit of making a gall on the leaf margin by causing an overgrowing and incurling in places, and since hardly a tree escapes its attacks, these marginal galls help the non-botanical person very much in separating this tree from others (Fig. 42). Belonging to the true laurel family, it has the fragrant leaves (Fig. 33) so characteristic of that group. It is a true laurel (*Lauraceae,* not the mountain laurel family), and is related to the tree from which the material is obtained which gives the characteristic fragrance to the bay rum of the barber shop. One thus has another

FIG. 42. THE LEAVES OF THE RED BAY ARE MADE RAGGED BY THE EVER-PRESENT APHID GALLS ON THEM.

FIG. 43. GALLBERRY (RIGHT) AND SHINING GALLBERRY (LEFT), BOTH NEAR RELATIVES OF THE CHRISTMAS HOLLY.

FIG. 44. STAGGER BUSH GROWS WELL IN MOIST UPLANDS. IT IS VERY ATTRACTIVE BUT ITS FOLIAGE IS POISONOUS TO STOCK.

check on the red bays by crushing the leaves and testing for fragrance. The flowers are not conspicuous.

The tree is not very desirable in plantings because of the unsightly galls which inevitably appear.

SHINING OR SWAMP GALLBERRY *(Ilex lucida)*. With its broad and very shiny leaves this close relative of the gallberry already described, makes an unusual appeal to the shrub-lover (Fig. 43). The leaf sketch brings out its peculiar character of widely spaced tiny teeth on the margin (Fig. 33). This species, unlike its near relation, prefers its soil very wet, and thus it becomes less available as a cultivated shrub. We have never seen this very attractive plant in cultivation, but if one has some low moist ground open for plantings it would be most worthy of trial.

STAGGER BUSH *(Pieris mariana)*. This shrub is so named because it is poisonous to stock. Preferring the drier bogs, it may also be found in sand where there is a shallow water table (Fig. 44). In flowers and leaves (Fig. 33) this plant resembles many others, but the dry

FIG. 45. THE ANCESTRAL TYPE OF ALL OUR FLOWERS, IS ILLUSTRATED BY THE FLOWER OF THE SWEET BAY. NOTE PRIMITIVE CONE OF CARPELS IN CENTER.

fruits which remain on the stems for a long time are helpful in identification. If there is little chance of its leaves being eaten, this shrub would be valuable in plantings. It should do well in medium moist upland sites.

SWEET BAY *(Magnolia Virginiana).* Probably the best known of all the shrub bog plants is this species of magnolia. Though commonly appearing as a shrub, it may, if absence of fire permits, grow into a tree of considerable size. Its familiarity to local people is related to its showy large leaves (Fig. 33) which, having the under sides brightened with a blue-gray bloom, may be seen and identified a long distance off. Again, the flowers (Fig. 45) are large and super-fragrant, by which we mean that if one smells them too closely and too long they become repugnant. In milder doses the odor is very pleasant.

It is of passing interest to learn that the magnolias are among the most primitive of our living flowering plants. The earliest true flowers were unquestionably like those of the sweet bay, and from these simple tree-ancestors have developed, according to our leading authorities, all of the smaller flower-bearing plants from which have been drawn the widely varied beautiful things of our gardens. The magnolia thus can be called a "living fossil," for it has come down little changed from the original type which was the basis for a vast differentiation into our present world population of one hundred and thirty thousand species of flowering plants.

Whether it may be correlated with this primitive character or not, the fact must be mentioned that the sweet bay is remarkably resistant to fire. When killed to the ground it sends up shoots, and if these are killed new ones come on and so indefinitely. The writer knows of trees which have survived for sixty years with fire destroying the shoots regularly every year during that time.

The sweet bay lends itself very well to cultivation in upland places. It will add a note of interest anywhere.

Titi or Leather-Wood *(Cyrilla racemiflora).* On the wetter bogs this ubiquitous shrub is almost invariably dominant. It is so wet in its preference that it forms the chief connecting link between the shrub-bog and swamp forest. The titi, if not set back by fire, may become a small tree when the more prominent trunk becomes conspicuous because of its very smooth light colored bark (Fig. 46). The leaf is wedge-shaped and rounded outwardly and possesses a characteristic veining which once being recognized proves helpful later (Fig. 33). The very small white flowers are borne in slender racemes which radiate in a striking fashion from the stem node, at the base of the new season's growth (Fig. 47). The fruits are small, dry capsules, and since these persist for some time, being borne in the same way as the flowers, they help later in the season in making an accurate identification.

Curiously enough the titi is only partly evergreen; i. e., the leaves remain on until December. While only present a part of the winter, it must be mentioned that they do much while they last to make the bog landscape rich in color. A trip during late October or November through that great shrub peat bog in Carteret County (the Open Grounds) will be more than worth while, for there the visitor will see a veritable riot of the rich reds of Cyrilla bushes scattered amid the deep greens of the other shrubbery. For the nature lover, the sight of a late fall titi bog, is one of the very finest North Carolina presents.

It is a pretty problem to understand how the roots of titi get their oxygen, when the water table is high for very long periods, which, with this shrub, is usually the case. The wood is very hard and compact so that gas diffusion through it apparently would be very slow. It may eventually be shown, as in the case of certain willows, that this plant carries on what is called intramolecular respiration, a type of energy release which goes on without oxygen. A very great deal is

FIG. 46. THE SMOOTH BARKED TRUNKS OF THE TITI ARE EASILY RECOGNIZABLE.

FIG. 47. THE TITI IN FLOWER; NOTE GROUPS OF SLENDER RACEMES.

yet to be learned about the life of these trees and shrubs which have chosen water-logged soils for their homes.

Because of its thin bark, titi is easily injured by fire. However, unless the base of the plant is destroyed (which is not usually the case), regeneration by new basal shoots takes place profusely and the taller stages soon reappear.

WAX-MYRTLE *(Myrica cerifera* and *M. Carolinense).* The fragrant leaves of this widely known shrub which ranges far north to Nova Scotia, constitute perhaps its most interesting feature. The gray-white wax-coated berries in an earlier day furnished the material for candles. The leaves (Fig. 33) are not unlike those of the gallberry in shape, but they are thinner and, in the second species named above, may be much larger. Very old ones assume a tree aspect (Fig. 40).

The wax-myrtle grows in the drier bogs and even ranges out on the upland where the water table is not too far beneath. It may be easily grown in cultivation where it should prove a most desirable shrub for a wide range of purposes.

WHITE OZIER (*Leucothoe elongata,* the species with erect flowers). This species is not a very important one, though occasionally seen in small societies growing in close or even pure stands. The branched terminal inflorescence with its numerous small erect flowers in racemes is very characteristic. The leaves are finely toothed (Fig. 33). The plant prefers the somewhat drier bogs, being thus commonly associated with gallberry. It is not nearly so widespread as the latter. Unlike the gallberry, this leucothoe is not evergreen.

FIG. 48. PENDENT FLOWERED WHITE OZIER IS CERTAIN TO ATTRACT THE NOTICE OF THE BOG TRAVELER.

FIG. 49. WICKY, THE BOG COUSIN OF THE MOUNTAIN LAUREL, A POISONOUS LOW SHRUB.

WHITE OZIER (*Leucothoe racemosa,* the species with pendent flowers). Even more frequent than the former is this interesting shrub, with its long horizontal racemes bearing the large pendent blueberry-like flowers (Fig. 48). The leaves are large with finely toothed margins (Fig. 33).

WHITE WOOD *(Xolisma foliosiflora).* An inconspicuous small shrub is to be found in this plant, one little desirable for use in plantings. It may be recognized by the presence of leafy bracts scattered through the rather compact terminal inflorescence. It prefers the medium wet localities.

WHITE WICKY *(Kalmia cuneata).* Also known as sheep laurel, this low shrub, a near relative of the mountain laurel, is widespread in our bogs, where it assumes a very modest rôle. If it is in flower anyone who knows the mountain laurel will instantly identify it (Fig. 49). Though the flowers are very similar they are arranged differently, for in the wicky they are borne in a strict raceme forming a cylindrical inflorescence very characteristic of this species. The leaves are narrow with rounded ends (Fig. 33).

Its poisonous properties are well known to stockmen. Horticulturally the plant has little value.

CHAPTER VI

THE MOST BEAUTIFUL GARDENS: THE GRASS-SEDGE BOGS OR SAVANNAH LANDS

On the eastern flat lands, Nature attains in our state her fullest and most varied expression of loveliness in the form of wild flowers. Only the spring display in the mountain meadows of the far west can rival the unusual wild flower show of our lower coastal plain grass-sedge bogs. And fortunately our bog beauties are not confined to the spring season at all, but during the summer and even in the late fall they continue to make glorious the sunny savannahs. The earliest settlers of the eastern lands, noted these areas and loved them for their beauty. John Brickell writing in 1737 describes them as follows: "Pleasant and delightful Savannahs or Meddows with their Green Liveries interwoven with various kinds of beautiful and most glorious Color and fragrant Odours which several Seasons afford. They appear at a distance like so many Pleasure Gardens being intermixed with variety of Spontaneous Flowers of various colours such as the Tulip, Trumpet-flower and Princess-feather."

What a strange relation it is that our most profuse production of wild flowers should occur from a sticky, black soil which when wet (as it is much of the time) presents a most uninviting appearance (Fig. 50). In

FIG. 50. A SOIL LAYER CAKE: THE HIGH WATER TABLE OF THE SAVANNAH PREVENTS DEEP ROOT PENETRATION AND RAPID DECAY OF ORGANIC MATTER, THUS FORMING A BLACK STRATUM, 1 FOOT THICK.

addition, if the flowers are to be abundant, fire must annually destroy the old season's debris, after which the burned areas are pictures of utter desolation. Such conditions characterize very many thousands of upland acres in the southern half of the coastal plain, where the level topography and non-draining subsoil conspire to keep a high water table in rainy periods. These grass-sedge bog soils are always of a mineral soil type; the peat soils seldom show a tendency to go into grass cover, but are usually blanketed by the evergreen shrub community. Naturally some bogs under the same rainfall frequency and intensity are wetter than others, so that we may distinguish between true or full bogs and semi-bogs. Curiously enough, it is the wettest bog which gives us the most flowers. Such

FIG. 51. A GENERAL VIEW OF A SAVANNAH OR GRASS-SEDGE BOG IN MID-SUMMER. THERE ARE NO "KEEP OFF THE GRASS" SIGNS ON THIS LAWN.

a one is the famous area near Burgaw, in Pender County, comprising fifteen hundred acres (Fig. 51), a local district which should some day be made into a state park.

A SEASON WITH THE SAVANNAH FLOWERS

We cannot do better in our attempt to describe the plant life of a savannah, than to take the reader on a seasonal excursion at the Burgaw bog, so that he may experience for himself, though indirectly, something of the impact such a remarkable succession of floral beauty makes upon the senses. He will then almost certainly be interested to go further into the chapter where the fascinating problems are raised which are concerned with the nature of the habitat capable of producing so unique a vegetation. If a gardener, he

FIG. 52. IN THE WINTER THE BURNED AREA OF A SAVANNAH HOLDS NO PROMISE OF THE FLORAL RICHES OF THE SUMMER.

might become interested in the project of making a small artificial bog in the yard at home.

Our seasonal excursion begins with late February and March, bringing a few soft spring-like days, and behold—what was a blackened waste from the winter's fires (Fig. 52) has changed its color to a delicate green tint. The very grass leaves which had been burned to the ground begin to grow at the base, below the ground, and soon an inch or more of light green leaves is evident. The miraculous procession of life has started.

Accompanying this primal greening, there develops the first flower of the season. Its local name (if it has one) is unknown to the writer, but if you will, in imagination, modify an ordinary dandelion plant as follows, you will recognize this interesting species: fill out the

FIG. 53. THE NIGHT-NODDING BOG DANDELION SLEEPS UNTIL 8:00 A.M. (LEFT), BUT IS FULLY AWAKE BY 10:00 (RIGHT).

leaves so that they are smoothly long-elliptical in outline, whiten the undersides with a close mat of white fuzz, change the "flower" (head of many tiny flowers) color to purple outside and white within, and then make the flower nod its head at night—and you have the night-nodding bog dandelion *(Thyrsanthema semiflosculare)*, the harbinger-of-spring on the savannah (Fig. 53).

With slowly increasing movement, the plants push upward from the ground. Myriads of primula-leaved violets *(Viola primulaefolia)*, conspire to form bright patches here and there. A little later, and appearing in a most unique manner because the flower comes out of the ground where no leafy evidence of the plant may be seen, the huge yellow blossoms of the trumpets *(Sarracenia flava)* loom up with their long petals waving languidly in the breeze. The scene of golden goblets scattered over a vast green festal board would not compare with that of the trumpet flowers dotted over the lawn-like expanse of the early spring savannah.

FIG. 54. THE BLUE BUTTERWORTS FILL THE PLACE OF THE ABSENTEE BLUE VIOLETS ON THE SAVANNAHS.

It is now the first of April. Already the blue butterworts *(Pinguicula pumila),* (Fig. 54) are raising their mottled throats to the sun, and the first delicate bladderworts *(Utricularia subulata)* are coyly pushing aloft their tiny yellow flowers on slender, hair-like stems. By this time also, tightly pressed against the dark earth everywhere, are thousands of the dark-red rosettes of the sundew *(Drosera rotundifolia),* minute clusters of glistening leaves which in themselves possess the beauty of a flower.

By the middle of April the vast expanse, despite the fact of the green background of grasses and sedges crowding in, takes on a white aspect. This is due to the appearance of the heads of a small-flowered fleabane *(Erigeron vernus),* which swarm up and out on

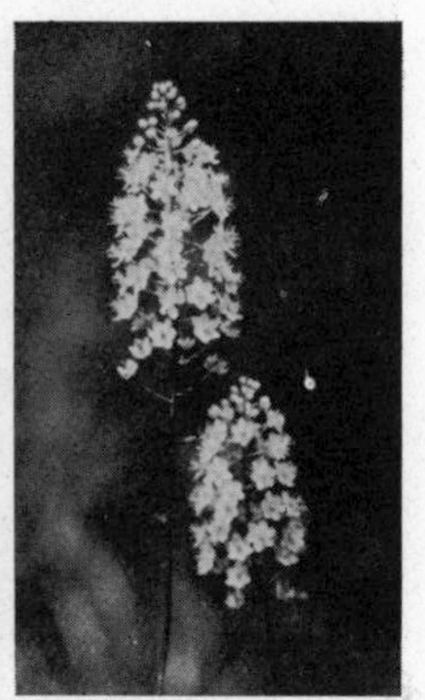

FIG. 55. SPRING BRINGS THE BOG FLY-POISON FESTIVAL ON THE SAVANNAH.

the area in enormous numbers. So light an aspect is this that the golden goblets of the late trumpet flowers are now spread on a white table cloth. In addition, among them may be seen a new course, in the red wine-filled glasses of the pitcher-plant *(Sarracenia purpurea)* flowers, while between these, everywhere over the table, the cloth has magically acquired myriads of blue polka-dots, in the appearance of the flowers of the blue-eyed grasses *(Sisyrinchium sp.)*.

The days come and go, and it is soon discovered that the banquet-table aspect is truly a fleeting one, for the time soon arrives at the end of April when the golden goblets and wine-glasses are removed and the white cloth fades.

Now a scintillating scene ensues in the panorama. Thousands of the hyacinth-like inflorescences of the eastern fly-poison (*Tracyanthus angustifolius,* fig. 55), together with the spire-like ones of the star-grass (*Aletris farinosa,* fig. 142), march out of the invisible onto the field. The small white flowers are borne on tall wand-like stems, so that they are always in motion,

FIG. 56. THE LOW SNAKE-MOUTH ORCHID (*Pogonia ophioglossoides*, LEFT) AND THE LARGE FLOWERED GRASS PINK (*Limodorum pulchellus*, RIGHT) WOULD NOT BE OUT OF PLACE IN A FLORIST'S WINDOW.

swaying and dancing with the full gayety of Wordsworth's daffodils. This scene certainly constitutes one of the main floats in the bog procession.

In early May some élite floral personages appear, and it must be remarked that for such aristocrats of the plant society as these, they appear in very large numbers indeed. The orchids now introduce themselves.

The small-flowered pale lavender pogonias *(Pogonia ophioglossoides)* may be seen by hundreds, each holding out laterally its remarkable crested lip (Fig. 56), the landing place of the insect aviator who carries the baggage of pollen. A few large-flowered pogonias (*Pogonia divaricata*, fig. 146) may be seen on the savannah margin where they surprise one with the unusual length of the lip, and the long waving narrow petals. Down amid the luxuriant grass, the tiny cream-colored spirals of a small lady's tresses *(Ibidium praecox)* may be found, along with the pale grass-pink orchid (*Limodorum graminifolium*, fig. 56). But most prominent of all is the queenly large-flowered grass-pink *(Limodorum tuberosum)*, an orchid which would attract attention in a Fifth Avenue flower shop. Varying from deep rose to purest white with a patch of yellow in the erect lip, they constitute one of the finest wild flowers of the entire country.

A garden would hardly be a garden without iris. And so it happens that middle and late May brings on the iris show. There is but one species (*Iris tripetala*, fig. 144), but it occurs in such numbers that locally the savannah is blue with its flowers, making the iris scene a brilliant display. The white delicately striped flowers of the Venus' fly-trap are now locally abundant.

Whole troops of the early orange polygala, or red hot poker *(Polygala lutea)*, begin to crowd in on the margins, soon followed by the flat-topped or branching polygala *(Polygala ramosa)* of a lemon-yellow color. Yellow star grass *(Hypoxis micrantha)* is seen everywhere. By this time many grasses and sedges come into flower, adding a minor note to the colorful ensemble.

With June come the pipe-worts *(Eriocaulon)*, small and large ones, appearing like magnified hat pins stuck in pin cushions. The modest bunch flower *(Triantha racemosa)* may now be seen everywhere, and more modest still, the sky-blue little lobelia flowers *(Lobelia Nuttallii)* are struggling to keep above the grass so that they may reflect the light of the heavens. But the great event which June brings in her train is the coming in of the meadow-beauties, two or three species of them. The largest is a tall herb (*Rhexia*

FIG. 57. TWO SPECIES OF MEADOW-BEAUTY, *Rhexia Alifanus* (LEFT), *R. Mariana* (RIGHT).

Alifanus, fig. 57) with a rose-colored flower measuring two to two and one-half inches across. One easily forgets the fierce shimmering heat of the midsummer savannah while he walks amid the innumerable splashes of light purple made by these large meadow-beauties.

Any day now the exquisite eight or ten-petaled sabbatia *(Sabbatia dodecandra)* may be met (Fig. 58). In this flower we have undoubtedly one of nature's sports or mutations, for the ordinary number of petals in the genus Sabbatia is four or five. At some time in the distant past a branch of the Sabbatia group was started with double the number of floral organs. And fortunate it was, for, without question, these showy larger flowers with their delicate rose-tinted petals are one of the very finest products of this natural garden.

Accompanying the Sabbatia, one may find on the savannah a milk-weed *(Acerates floridana)* giving out

FIG. 58. THE EIGHT TO TEN PETALED SABBATIA SUGGESTS A ROSE-COLORED DAISY.

FIG. 59. A "FLOWER" NOT A FLOWER: THE WHITE BRACTED SEDGE.

a fragrance delicate but distinctive. Here and there in patches are regiments of white-topped sedges (*Dichromena latifolia,* fig. 59), the pure white bracts of which so perfectly simulate flowers that every non-

FIG. 60. THE GREATEST DISPLAY IN THE RICH COASTAL PLAIN IS THE FOURTH OF JULY PICNIC OF THE SAVANNAH WHITE ORCHIDS.

botanical person would declare they were the showiest of blossoms. And everywhere now, down in the grass, the curious light green fine-leaved running pine *(Lycopodium alopecuroides)* is creeping and branching, furnishing an excellent "filler" for a June bouquet.

Most of the June attractions hold on into July and some into August. Yet, in addition, July brings on some features of its own. The savannah lilies *(Lilium Catesbaei)* silently take their place at well scattered stations, there to spread their slender spotted orange petals to the shining skies. One flower or sometimes two are borne erect on the tip of the stalk. They stand well apart like sentinels or, shall we say, traffic cops, in the great flower parade.

Now appears perhaps the finest display of all. Of

those attractions that have come and gone before, one may see the procession from any point on the savannah, but to see the massed display of floral loveliness about to be mentioned one must go to the northeastern end of the area near the railroad, for the plants in question have not yet spread over the whole garden. Here, standing in places so thick that they touch each other like people in a crowd, one may see in favorable years thousands of the white savannah orchids (*Gymnadeniopsis nivea,* fig. 60). Each plant bears a mass of the purest white spurred flowers, and they all apparently come on the scene at about the same time. When at its height this display of wild flowers is positively one of the finest in America. The writer personally knows of nothing to match it. It is worth going miles to see. The best time is between the first and the fifteenth of July.

In August the tallest species with showy flowers comes in, the tall bog bunch flower *(Zygadenus glaberrimus),* four feet high with its lax cluster of prominent star-like blossoms (Fig. 61). They may be picked out easily at long distances across the savannah. With them one may now find the radiant marshallia *(Marshallia Williamsonii)* and exquisite flat-topped umbels of oxypolis *(Oxypolis filiformis),* and the curious, coarse, almost ugly, hairy massed flowers of the Carolina redroot *(Gyrotheca tinctoria).* Two interesting eryngiums *(Eryngium integrifolium, E. aquaticum)* should also be mentioned, but their head-like inflorescences are not very conspicuous. The slender stemmed gerardias may now be seen, as August closes, beginning to splash purple masses over the area. They make a floral chain binding August to September.

FIG. 61. THE GLANDS AT THE BASE OF THE SEPALS AND PETALS ON THE TALL BOG BUNCH FLOWER ARE HELPFUL IN RECOGNIZING THIS PLANT.

With September, the turn of the season begins. The savannah reflects the change ever so inconsequentially at first, but by the end of the month the presence of the blazing stars *(Laciniaria)*, the pleeas (*Pleea tenuifolia*, fig. 62) and tofieldias *(Tofieldia glabra)* tell the story of the approaching end of the procession.

From the late summer aspect of deep green grass spotted with multicolored flowers we now observe the savannah as October comes in, bringing on the final gay plant masses. To attain this golden color of the final scene, the rayless flat-topped goldenrod *(Chondrophora nudata)* is used. It is a close relative of the goldenrod, but instead of having its flowers in a more or less vertical mass, they are arranged in flat-topped clusters like those of Queen Anne's lace. Countless myriads of them appear, resulting in the savannah developing

FIG. 62. THE STAR-FLOWER MAKES CERTAIN AREAS AS BRILLIANT AS THE NIGHT SKY.

a sort of golden-flowered second story. Other yellow elements enhance the effect of Chondrophora in putting on the final gorgeous scene. Coreopsis *(Coreopsis falcata)*, unusual species of sunflowers, and the savannah golden-rod *(Solidago salicina)* all come in to make the grand finale the thrilling golden spectacle it is.

Late November comes. The pageant is over. Yet nature has furnished an anti-climax in the asters *(Aster paludosus, A. elodes)*, which in their cool blue tints suggest the coming of the frosts. Even December still presents the asters, the last offering of the savannah gods. January alone has not a flower to brighten its passing.

OTHER SAVANNAH SPECIALTIES

In the foregoing account of a year on the savannah, we were so interested in the flowers and their beauty that we lost sight of a few peculiarities of this vegetation which are of especial interest. Casual examination of the leaves discloses the fact that they are for the most part narrow and small, and also possess the peculiarity of being hard and tough. Cattle can feed on the grass only in the spring. Further, on dozens of our plants with the showy flowers, the leaves have the habit of standing erect with their upper sides pressed against the stem. For the probable explanation of certain leaf modifications which are consistently true for an entire community of over a hundred species, we must look to the soil habitat.

Before taking up the discussion of the bog soil, brief mention must be made of the grasses and sedges which are so abundant as to force us to call our community a "grass-sedge" bog rather than to name it after some of the gorgeous flowers which are scattered over it.

Of preëminent importance is the orange-grass (*Campulosus aromaticus,* fig. 134). It may be instantly recognized by its scythe-shaped compact head of spikelets. These latter, the little units of the inflorescence, are borne on one side of the connecting stem and each one bears a few delicate bristles. A final test is the odor. Shut the eyes and crush the fresh head and you will imagine yourself in the orange groves of Florida. Curiously enough, the rest of the plant does not produce the odor.

The other bog grasses play an insignificant rôle. Among these the bog panic grasses are most frequently seen. The sedges are less prominent than the grasses, but the nut sedges, with their curious white seeds resembling little stones, are very frequently observed.

Though not dominants by any means like the orange-grass, the various species of pitcher-plants, or trumpets, are so distinctive of savannahs and so inter-

FIG. 63. THE INSECT CATCHING TRUMPET LEAVES IN RED AND GOLD ARE AS SHOWY AS FLOWERS.

esting in themselves that some account of them must be introduced here.

The well known habit of these specialized leaves (Fig. 63) which contain free water is that of drowning insects and, by means of enzymes secreted into the water, digesting the bodies and absorbing the soluble nitrogenous material for food. The pitcher plant *(Sarracenia purpurea)* is unprotected from overflow in the rains with resultant loss of its meal. The other two, called trumpets, are never seen full of water because of the umbrella and the rain hat respectively. The water is secreted in their cavities. Young trumpet leaves which have not yet opened their mouths in their growth may be observed to contain half a cavity of water.

The insects are attracted by odor, which gains in

FIG. 64. NO SHOWER CAN DISTURB THESE LEAF INSECT TRAPS OF THE RAIN-HAT TRUMPET. NOTE "WINDOWS" OR CLEAR SPOTS ON THE BACK SIDE.

intensity as the dead bodies of the bugs increase in number. At the periphery, on the upper side of the trumpet hood, a series of nectar bearing areas may be observed which are attractive to insects. In addition, the leaves themselves are colored bright yellow in *Sarracenia flava* with more or less development of red along the veins, and this may be important in catching the attention of the small creatures of the air.

The rain-hat trumpet *(Sarracenia minor)* is a most amazing plant (Fig. 64). If one holds one of these leaves in front of a light and peeks under the hood cover, he will see that the white patches on the back are very transparent. It may well be true that insects attracted by odor will go on in because of the light coming through the windows. Once in, however, they

FIG. 65. WHERE ANTS AND THISTLE ROOTS LIVE TOGETHER. SPRING AND SUMMER ASPECTS OF LECONTE'S THISTLE WHICH ON THE WETTER SAVANNAHS IS WHOLLY CONFINED TO THE ANT HILLS.

cannot easily escape since the smooth sides prevent egress by crawling out. A buzz or two and the insect is helplessly drowned in the water.

Not only because of their remarkable food habits but for aesthetic reasons as well every yard bog should have all of our native insectivorous plants of this kind. They are extremely graceful in form.

The absence of certain plants from an area is often as striking a feature as the presence of others. On our true bogs, leguminous species are never seen. And for some curious reason the largest genus of sedges *(Carex)* is not represented. However, on the semi-bogs where drainage is better, both of these are found.

One exceptionally interesting minor situation in the presence and absence of a plant is that involving Le Conte's thistle *(Carduus LeContei)*. On the wetter savannahs this plant is strictly confined to the ant hills of the savannah ant (Fig. 65). These little chocolate drop-shaped mounds are 4-6 inches high and remain above high water during the rainy seasons. In the summer when the grass hides the ant's domiciles, they may nevertheless be easily located. The tall, showy heads of thistle flowers standing high above the grass, will disclose the hidden homes of the ants every time. Strangely enough, no other wild flower tries to contest the ant hill site with the thistle.

We shall hope that the reader by this time has been sufficiently intrigued by the floral glories of our grass-sedge bogs, to follow through the succeeding brief discussion of the special complex of conditions which brings such a botanical paradise into existence. It is truly a fascinating story.

We have already discussed the bog habitat in the preceding chapter on shrub bogs. Just remember that on bogs you may merely get your feet wet, while in marshes and swamps you must generally wade. The writer once stood in a heavy downpour on the Burgaw savannah and watched the thin sheet of surface water flow rapidly around his feet as it quickly found its way to the slightly lower swamp at the margin of the area. Thus, so far as the two factors of topography and non-draining subsoil go, our herb covered areas are exactly the same as the shrub covered ones. Hence, the question immediately arises: Why are the communities different?

Fire Helps Make the Savannah

Occasionally in this curious world of ours, destructive agents can carry out constructive activity as well. Our beautiful savannahs constitute an extraordinary example of this paradox, for they are dependent for their existence and perfection upon fire. Without the aid of this scourge and destroyer our humble herbs would have no chance whatever in competition with the shade-making bog shrubs and pocosin pines. By the simple device of placing their stems under ground out of reach of the flames, the perennial herbs come through the conflagrations without "the loss of a man." The shrubs, however, are set back terribly when severe fire burns them to the ground. Many will be killed outright while others will be so weakened that regeneration by basal shoots takes place slowly. Thus it is that on the mineral soil bog areas of our coastal country there has been a movement related to fire frequency, from the shrub to the herb type—the pocosin becomes a savannah. Intermediate stages are still abundant in which can be uniformly read this story of death and survival on the part of two distinct vegetations occupying the same soil habitat.

We must not pass by this point in our narrative without pausing to pay a compliment to two trees which are able to survive indefinite periods of annual fires. Though not abundant these relict trees may be found on old savannahs which have certainly been in existence for fifty or more years. In both instances a few thick trunk-like branches radiate from a common taproot and being half buried in the ground only suffer injury on their upper sides. At the ends the shoots appear regularly every year. These trees, the swamp gum (*Nyssa biflora,* fig. 66) and sweet bay *(Magnolia Virginiana),* apparently are the only woody plants which can successfully imitate the perennial herbs by partially burying their stems, to make possible survival. In the summer the numerous shoots arising from the ground make them appear as shrubs, but few

FIG. 66. A TREE DWARF: A GUM TREE *(Nyssa biflora)* WHICH HAS SURVIVED 75 YEARS OF ANNUAL FIRES. NOTE SHORT TRUNKS WHICH SPREAD AT GROUND SURFACE BEARING THE SHOOTS OF THE SEASON.

observers would ever imagine these "shrubs" to be trees fifty, eighty, or a hundred years old.

SOIL CHARACTERISTICS

All that was stated about the nature of the shrub bog water-logged soil is equally true of our grass-sedge bog substratum. The low oxygen interfering with root growth and its function in water absorption is to be correlated with the reduced and appressed leaves of the plants. The unpalatable summer condition of the grasses making savannahs poor grazing areas is due to the same fundamental cause. Like the bog shrubs our savannah herbs are using ammonia instead of nitrates, since the latter are totally absent.

Of many curious features which the writer and his colleague, Dr. I. V. Shunk, discovered about the savannah bogs, none was more baffling than the fact that when the drainage is improved by a road ditch, for example, the savannah plants hold on to the area right up to the ditch edge. Even when these plants were dug up and every chance given for weeds to enter the disturbed area they could not do it. Evidently the long series of periods of flooding have so permanently changed the soil in some way, that when the flooding ceases other plants cannot get started.

In closing our discussion of the bog proper a brief statement should be made concerning our use of the term "bog." Many authors confine its use to peat areas, but since the water-logged mineral soil habitat is fundamentally the same as that of the soft organic matter, and since many plants such as the sundew, pitcher plant, and many flowering herbs (orchids particularly) are the same species as found on the northern quaking bogs, it seems logical and desirable to broaden the term to include the solid mineral soil areas of the south.

Attention should be called to the fact that very restricted areas both in the piedmont and mountains are characterized by bog conditions and carry typical bog plants. Many of the southern savannah species have found their way into these piedmont and mountain bogs.

The Semi-Bogs

Like political parties, our bogs differ in degree of wetness. Those which are better drained due to the presence of coarser sand in the soil or the introduction of slope making possible more rapid lateral movement, may well be called the semi-bogs. These may easily be distinguished by the fact that the dominant grass over them is something other than the orange-grass; yet many of the characteristic true bog wild flowers, together with the insectivorous trumpets and pitcher plants, will still occur to tell the story of the bog condition. The principal grasses found on the better drained bogs are the scaly rootstalked panic grass *(Panicum virgatum* var. *Cubense),* the semi-bog broomsedges *(Schizachyrium scoparium, Andropogon capillipes)* and strangest of all the sandhill wire-grass *(Aristida stricta.)*

When a very common herb like the sandhill wire-grass is found living in another and apparently entirely different habitat, the ecologist must take notice. It affects him much as if he found a sparrow diving into water and amusing itself as would a duck. For the upland sandhill wire-grass, this is simply not an orthodox thing to do. Yet the wire-grass makes that great a jump in its habitat relations. In its community as described in the chapter on the sandhills it has its particular set of associates, yet in the semi-bog it leaves all of those herbs behind and is found in entirely different company. Only the long leaf pine can go with it in its jump into the new habitat.

Two other wire-grasses *(Aristida palustris* and *A. lanosa)* are to be found in perennially wet areas, but these never are seen in the high and dry sandhills. Only the ubiquitous *A. stricta* seems to tolerate this wide range in soil water conditions.

Such unusual versatility on the part of a plant must

have some correlated peculiarity in the habitat. From the restricted observations so far made it seems that the explanation in some areas is to be found in the occurrence of a strange kind of hard-pan formed from organic matter, lying 14-20 inches below the surface. Such a layer is totally impervious to water and even roots cannot penetrate it. Such a subsoil structure readily maintains flooding during frequent rains, while during drouths the soil proper above the "ortstein layer," as it is called, will become very dry. Only the wire-grass and certain bog herbs among the low plants can stand such alternating extremes in the soil water condition; hence their survival there. The flooding of some wire-grass for ten weeks failed to injure it.

This soil habitat is a notable one for the reason that, for a soil of such a depth, it probably goes through the widest swing in its water content of any soil in the state. So when the reader finds himself on an area covered by wire-grass and certain bog wild flowers (for all cannot stand the dry seasons), let him feel a sense of high regard for plants which can survive over long periods such an extreme change in the amount of soil water present.

Before leaving the discussion of the semi- or better-drained bogs, mention should be made of the best drained flat areas in the low country, which, under the condition of frequent fire, carry a broom-sedge *(Andropogon* or *Schizachyrium)* cover. With the loblolly and long leaf pines which may occur in open stand amid the luxuriant broom-sedge grasses, these areas present most picturesque types of savannahs (grassland with trees) and add much to the distinctive character of our coastal lands. Since the sandy soil of such

FIG. 67. "THE MOST WONDERFUL PLANT IN THE WORLD." A GROUP OF VENUS' FLY-TRAPS READY FOR INSECT VICTIMS.

areas is of a medium texture, together with other conditions making possible good drainage, the broom-sedge savannahs may not be classified in the wet land or bog series. Indeed, the broom-sedges on such areas plainly tell the land buyer that here is favorably watered soil, one that does not become, under average weather conditions, excessively wet or excessively dry. Among the trees of the region the loblolly pine tells a similar story. The late W. W. Ashe, well known North Carolina botanist and forester, long ago pointed out the close companionship which broom-sedge and loblolly pine held, in relation to habitat; they have always lived together in the same soil home, a fact equally true whether the soil is a coastal sandy type or a piedmont clay slope type.

The Most Wonderful Plant in the World

On those moist pocosin borders or flat semi-bog areas in the southeastern corner of North Carolina, where the soil does not become excessively wet or dry, is to be found the most unusual of unusual green things. This is the plant which Darwin called "the most wonderful plant in the world," the Venus' fly-trap *(Dionaea muscipula,* fig. 67). And the modern botanist who has learned much about the plant which Darwin did not know, is quick to agree with him. In this plant the principle of the common steel trap is seen to have long antedated that of invention; indeed, the fly-trap long antedated man himself.

Not only the great student of evolution recognized the unique character of this plant but earlier, in 1768, the noted Linnaeus wrote the following letter to John Ellis of London, who had sent him specimens: "I yesterday received your welcome letter accompanying the description, character and figure of that most rare and singular plant, the Dionaea, than which certainly nothing more interesting was ever seen. Though I have doubtless seen and examined no small number of plants, I must confess I never met with so wonderful a phenomenon."

It is of interest to note that our principal insectivorous plants, sundews, pitcher plants, and fly-trap, are closely related according to floral structure. Different as the leaves are, however, the distinction between petiole and blade is evident in all of them, so that the trap of the species under discussion, the hood of the pitcher plants, and the expanded ends of the sundew leaves are all comparable, or homologous, as the biologist puts it. In this group, nature has carried out the most amaz-

FIG. 68. EVEN A GRASSHOPPER (UPPER CENTER) CANNOT ESCAPE THE CLUTCHES OF THE FLY-TRAP.

ingly different modifications of leaf parts, to arrive at the same end; viz., the capture and digestion of insects.

The Venus' fly-trap leaf stalk is an expanded structure and, being green, appears leaf-like. The blade or trap portion, if exposed to the sun, is bright red, giving the plant a very showy aspect. It may be true, as some have suggested, that this color attracts insects to their doom. The immediate structure of most importance to the bug-life of the semi-bogs where these plants grow, is, of course, the triggers which occur in a group of three near the center of each half of the wide-flaring leaf trap. Oddly enough, the triggers must be touched twice; the trap is not to be sprung by a mere chance particle of inanimate matter.

How do the leaf halves snap shut and later press the victim tightly between (Fig. 68) them while the digestive juices dissolve away the soft parts? The researches indicate that the same kind of pressure in leaf cells which gives way when a leaf wilts and droops, is present. This osmotic pressure is used, however, not merely to keep the leaf distended as in most plants but to press the two sides open against a spring mechanism formed by the woody substance in the veins. Thus it is that when the stimulus from the hair trigger rapidly passes through the living cells of the upper side of the leaf, these cells, swollen with water under pressure, now rapidly allow this water to escape into the space between the cells. The tension released, the woody springs come into action, the trap snaps shut, and the plant world has a partial revenge for all of the vegetation that has been eaten by its minute animal enemies.

The flowers are very attractive (Fig. 69). They are all attached by short stalks to the end of the common tall main stalk. They bloom in late May, so that this is the best time to study distribution since the flower cluster is sufficiently showy and elevated to be seen from a distance. Strangely enough, the flower must be placed among the primitive types, ranking low in the scale of floral complexity.

Not the least of the many curious facts about the fly-trap is that of its highly restricted distribution. It is located frequently and in some abundance only within a radius of seventy-five miles of Wilmington, North Carolina. Beyond this range are scattered outlying colonies, and, according to Dr. W. C. Coker of the University of North Carolina, the most distant station from Wilmington is about one hundred and fifty miles west of that city, near Wadesboro, North Carolina.

FIG. 69. THE BLOSSOMS OF THE VENUS' FLY-TRAP ARE DELICATE AND ATTRACTIVE WILD FLOWERS.

The habitat is distinctly that type of semi-bog which does not get excessively wet (long periods of flooded soil) or excessively dry. Certain tall plants are commonly associated with it such as the round-leaved boneset, the hound's tongue or vanilla plant, and certain blazing stars. Such low plants as the running blueberry and the pyxie flower also occur with it.

Many botanists have the notion that it is a rare plant. Fortunately that is not true. There are many localities where it grows so abundantly that one cannot walk across the area without stepping on it. This marvelous plant is in no danger whatever of extinction.

CHAPTER VII

DESERTS IN THE RAIN: THE WIRE-GRASS SANDHILLS

AMID THE white sands of our sandhill areas the sun shines both up and down; here one can acquire a fashionable tan with one's hat on. Here the rain disappears as fast as it falls and the soil surface is dry within thirty minutes after the summer shower is over. Scattered here and there in patches, or even covering whole ridges, is the "granulated sugar," of the coarse, white, rain-washed sand reflecting light and heat like a mirror. Sterility from leaching, heat (with high evaporation), from reflection of the sun's rays and very low soil water in the long drouths and terrible fires all conspire to make the sandhills very different and unique localities for plants, animals, and men.

In such a special soil area we may observe the most striking cases of response of organisms to an unusual environment. Even the human species has reacted by developing in such a region one of the most fashionable winter resorts in the country. A curious paradox this, since only the poverty grasses and poor herbs can grow here in a state of nature. Nevertheless it is true, that the same combination of climatic and soil characteristics, which makes the sandhills so distinctive in habitat and vegetation, is the same which proves so attractive to the wealthy winter golfer and cold season vacationist. But there is one important consideration: the plants must remain through the summer, and that is another story. It is this story with which we are to be chiefly concerned.

The Story of Sandhill Plant Life

The reader will, of course, understand that when we are speaking of the sandhill vegetation we are not speaking geographically, for any large sandhill area will include between the hills much low ground carrying swamp forests and shrub bogs. These are treated elsewhere in their respective chapters. We are interested in the upland plant life. And to introduce the most distinctive vegetation, we must begin with that community of plants which is able to survive on deep coarse sand.

These species are the born "drys" of the North Carolina deep soils. Mere thin crusts of lichens can live on a granite rock surface (the driest habitat under the sun), but we are here concerned with soil rooted plants—the ones which under the condition of reduced capillary movement of soil water and other unfavorable conditions can still survive the long hot summers.

The type community of the deep coarse sand may readily be recognized by the presence on it of two trees and only two, the long leaf pine *(Pinus palustris)* and the forked-leaved black jack or turkey oak (*Quercus Catesbaei*, fig. 71). If any other trees are present, the field student will know immediately that some condition has changed, resulting in improvement of the soil water condition. Only the ecologist, or the student who is always studying his organisms in relation to their environment, fully realizes how accurately the

FIG. 70. THE "WETS" (POCOSIN) AND THE "DRYS" (SANDHILL PLANTS) IN CLOSE CONTACT. SCENE NEAR WHITE LAKE.

FIG. 71. TURKEY OAK AND LONG LEAF PINE, THE TWO TREES OF THE COARSE SANDS.

presence and absence of plants tell the story of soil conditions. How valuable the wild vegetation is, then, for the land buyer! He can read the story of the soil as it is recorded by the plants more plainly than in a book—if he has observing eyes.

To one who has given some study to the soil conditions of the coarse sands, it is an impressive sight to see the youthful pines and turkey oaks making the grade on one of those glistening white sand areas where millions of other seeds are deposited every year only to be followed by the early death of their seedlings. A few remarkable herbs may be observed surviving with the trees. These do not include wire-grass, for even this camel of the grass world is eliminated, so extreme is the Sahara of our white ridges.

Excellent examples of the very coarse sand ridges are to be observed between Wilmington and Carolina Beach. From the road which cuts through them they can be seen to be ancient bars, dating from an earlier age when the Cape Fear peninsula was under water. However, so common are such ridges, large and small, further inland in the southern half of the coastal plain (Fig. 72) that the tourist would have to be wholly city-minded not to notice them. Extensive areas of these local soil deserts are to be found near White Lake. They are so thickly dotted all over the middle and lower coastal plain, however, that enumeration would be impossible.

Before returning to the great sandhills, we should pause a moment to study these North Carolina desert-like plants which have adopted ways and means of meeting this, the driest and most sterile of our soil habitats.

FIG. 72. WIRE-GRASS SEMI-BOG AND DISTANT WHITE SAND RIDGE.

FIG. 73. THE YOUNG TURKEY OAKS TURN THEIR LEAF-BLADES EDGEWISE TO THE SUN.

The seedling turkey oaks uniformly twist the short leaf stalk sufficiently to orient the blade vertical to the sand surface (Fig. 73). Under the condition of high reflection, the gain of the leaf is readily apparent. When the writer and his colleague, Dr. Shunk, wired a series of leaves in the horizontal position, injury resulted. This vertical leaf orientation, the character of deep lobing (reducing the amount of water transpiring surface) and a thick waxy coat on the leaf are all adaptations enabling this broad-leaved tree to survive under the condition of very low water absorption during drouth periods. Thus the turkey oak takes its place as the "driest" of our wide-leaved trees.

The young long leaf pine, with its narrow needle-like leaf, having its leaf pores in grooves, begins its life with a structure favoring water conservation. It is of interest to note in addition that the seedling pine does not bend its leaves over to expose them broadside but holds them stiffly erect, thus escaping the direct incidence of the sun's rays.

Every naturalist is kept in a state of astonishment over the diverse ways by which organisms meet the same set of habitat conditions. Our pioneer herbs on the white sands furnish a striking illustration. These will, however, be discussed later in the chapter.

The Great Sandhills

Returning to the great sandhills of the southcentral region of the state we find this vast undeveloped area reduced through lumbering and fire to a state of extreme vegetational poverty. Like a fine old aristocracy destroyed by war, the original long leaf pine forest of our sandhills has been completely cut and

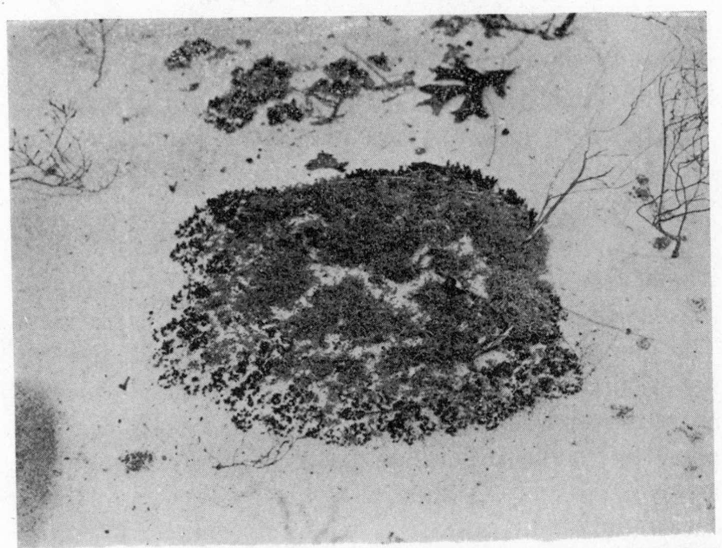

FIG. 74. THE SANDHILL SELAGINELLA FORMS A PARTNERSHIP WITH THE REINDEER LICHEN IN THE CENTER.

FIG. 75. LICHENS ON THE SAND. THEY ARE NOT ROOTED AT ALL, BEING READILY KICKED ABOUT.

burned until the entire scene has so changed that no person of the rising generation may now gain any real idea of the majesty and glory of the original forest.

The early settlers who came to the Sandhills found all of the uplands in long leaf pine, which they called "yellow pine." A report written in 1810 stated that "the Sandhills yield no other timber than yellow pine, which from the sameness and uniformity of appearance affords too little variety to be pleasing to the view. The gentle inequalities of the surface are all that relieves the wearied eye." It is to be suspected that this early writer was not as sensitive to the beauty of great monotony as he might have been. For any true nature lover of that day, it must have been a thrilling experience to have traversed the seemingly endless mazes of the virgin long leaf pine forest amid the sixty to one hundred feet high boles, under the shady tasseled tops whispering or roaring eternally in the winds. Underfoot was the deep brown carpet of needles, broken here and there by wire-grass tussocks. So compact was the needle-leaved canopy that trail travelers of that early day could go for miles in comfortable shade. On the level, deep sand hilltops the trees were of slender stature. On the lower slopes, or where the roots could make contact with the favorable water retaining clay or red sand, the giant trunks were found ranging around three feet in diameter. At no other point in its native Southland did this, the most beautiful pine in the world, attain a finer development than it did in our own Sandhills.

And now this noble original forest, one of nature's most unique products of the ages in North America, is gone—rooted out by hogs, mutilated to death by tur-

FIG. 76. THE WORK OF THE FIRE VILLAIN. JUVENILE LONG LEAF PINES ARE STILL ALIVE, HOWEVER.

pentining, cut down in lumbering, burned up through negligence (Fig. 76). No "Save the Pines League" was ever formed to rescue any of it. Not a part of this great natural wonder, worthy of the name forest, remains intact within the state's borders. White "civilization," just as it has in the North and the West, has triumphed over nature in the interest of "human welfare."

As the tourist in northern France must still look upon the ruins of the war, so the sandhills visitor of today must accept the glaringly hot, scrubby, fire-scorched scene of today (Fig. 77). Yet these areas are not without botanical interest, since the plants in them exhibit remarkable adaptations making possible their survival.

Following the destruction of the great pines, other

FIG. 77. A LONG LEAF PINE SILHOUETTE AMID THE WINTER TURKEY OAKS.

and smaller plants had their chance. The turkey oaks (called forked-leaf black jack, an awkward name) increased, since, once established from the acorns, they thereafter reproduced by basal shoots after being killed to the ground by fire. This behavior has resulted in most of the tree trunks today appearing in groups of two to four, these being regeneration shoots from a common root crown.

Among the few smaller plants which deserve special mention, none is more common or more remarkable than the wire-grass (*Aristida stricta,* fig. 78). This is the very common tussock grass with slender wire-like leaves, which one encounters everywhere in the coarse sand country. It not only grows on the hilltops but down the slopes as well, until in thick stand it makes

FIG. 78. THE CAMEL OF THE GRASS WORLD IN NORTH CAROLINA—WIRE-GRASS. NOTE THE FINE STIFFLY ERECT LEAVES.

contact with the shrub-bogs of the wet soils on the lower slopes and bottoms. Wire-grass may everywhere be used as the indicator of medium to coarse sand soils in our coastal plain. Wherever the sand runs toward the finer types it is never found. Omitting the very coarse pure white sand ridges, previously discussed, we may say that wire-grass and sandhills are like Siamese twins—never seen apart.

In habit, strange to relate, this grass is shallowly rooted where the sand is deep, most of the roots growing straight away horizontally in a zone one to six inches deep. In the longer drouths this layer dries out completely and in sand capillary forces cannot bring water from the deeper layers. Yet this plant survives the longest drouths. The secret lies apparently in the structure of the leaves which have so many woody cells

FIG. 79. READY TO BURN. THE WIRE-GRASS HAS HAD A FEW YEARS TO THICKEN THE TUSSOCKS. JUST ONE CASTAWAY CIGARETTE MAY NOW DEVASTATE A HUNDRED SQUARE MILES.

in them that they really are but stretched out toothpicks. The tiny leaf folds together along the middle, one edge overlapping the other, a device that does much to prevent the loss of water. Further, the amount of living tissue to be kept alive in the leaf is reduced to a minimum, the rest being dead wood. Thus in wire-grass we see a striking illustration of how plants meet low water conditions: they reduce the relative amount of living cells, and besides check as best they may the evaporation of water. Other plants use still different methods.

We may now easily understand how it is that wire-grass is inflammable. Every tussock, especially the old ones with many dead leaves in them (the wood resisting decay) is naught but a pile of the finest slivered kindling wood (Fig. 79). Just a few hours of a drying sun on wet wire-grass will get it ready for a conflagration. It may burn at any time of the year. To it may be largely ascribed the high fire incidence of sandhill country. No region anywhere presents a more difficult problem in fire control than does that of our wire-grass covered sandhills.

To local folk in the sandhill country, it is a familiar observation to note how fire stimulates the production of the slender fruiting stalks of the wire-grass. Where a road has passively functioned as a fire guard, one may ride along for miles in the fall of the year and see the grass fruiting on one side and not on the other, in relation to the presence or absence of preceding fire.

Before leaving the discussion of wire-grass, mention must be made of a mystery concerning it. Once it is plowed up, as it has been over thousands of acres, it will not return when the field is abandoned. Cotton patches and peach orchards abandoned over ten to twenty years ago show no trace of its coming back; yet it will be thick in the adjoining woodland right up to the old field edge. Here is a botanical "believe it or not" which needs investigation. Its place is taken by a thin stand of depauperate weeds, not the least important of which is the barefoot boy's enemy, the sand-bur.

A volume could be written about the long leaf pine *(Pinus palustris)*. There are a few facts, however, which everyone should know concerning this the noblest tree of its group (Fig. 80). Mature cones are not produced every season, seed years being from three to seven years apart. The seedling pines spend the first four or five years building a tap root, the stem not rising above the ground surface. The cluster of leaves

FIG. 80. THE FEATHER DUSTER FOLIAGE OF THE LONG LEAF PINE MAKES AN ATTRACTIVE PICTURE AGAINST THE SKY.

FIG. 81. WHAT A TREE THIS LONG LEAF PINE YOUNGSTER WOULD MAKE IF GIVEN A CHANCE!

FIG. 82. A BIG LONG LEAF PINE SAVED BY ISOLATION IN THE CENTER OF ANGOLA BAY, A GREAT PEAT BOG OF PENDER COUNTY.

during this stage resemble a coarse wire-grass. Fire will sweep over them destroying the leaves, but the stem bud will come through unscathed. Even the young trees (Fig. 81) because of rapid growth and early appearance of thick bark, will resist fire most remarkably. This pine is the king of fire resisters among the trees. However, especially heavy fires under high wind will kill many of them. Some degree of fire control should always be a consideration in the restocking of these long leaf pine sites. Studies by the United States Government forestry experts have shown that reforestation in the sandhills can be made a profitable enterprise with reasonable fire control. Our long leaf pine barrens should be artificially reseeded. In a few years their appearance could be completely

FIG. 83. THE SMALL LEAVED PYXIE OF THE HIGH SANDHILLS. FOUND ONLY NEAR SPOUT SPRINGS, N. C.

changed by the highly decorative foliage of myriads of the young trees.

Brief mention must be made of a very common little shrub which seems to supplant the wire-grass where the turkey-oak trees are thicker. We refer to the running huckleberry *(Gaylussacia dumosa),* a low, nearly evergreen shrub which spreads by underground branches, from which at intervals clusters of roots are given off. The berries are black but not very good to eat.

Of especial interest is a recently discovered species of pyxie or "flowering moss" (*Pyxidanthera brevifolia,* fig. 83), which is only known from a restricted area some six miles square in the sandhill district near Spout Springs in southern Harnett County. It had escaped detection until the author had the good fortune to find it in 1928.

It was named "brevifolia" because of the very minute size of the smaller leaves, these being commonly less than 1/16 of an inch long. Closely overlapped on the slender, creeping stems, they give the plant a distinct moss aspect which is much emphasized by the close and compact branching system, resulting in the formation of mats. In the region of greatest abundance, one mile south of Spout Springs on the Sanford-Fayetteville highway, these mats are frequently three to five feet in width. They are so common here within a short distance from the road that even the most casual observation will pick them up.

This new species differs from its only other relative, *P. barbulata,* in a number of ways: The leaf is 1/16-3/16 inches long, the flowers are smaller, and the plant assumes an aspect of a compact mat. The soil habitat is of deep, dry, coarse sand. Its distribution is highly restricted, the species at the present time being known only from the region in the vicinity of Spout Springs. The flowers bloom in midwinter (February).

Why and how such a species came to be so isolated would be difficult to explain. The seeds have been collected, but they were found impossible to germinate. Such a plant, apparently no longer reproducing by seed, has under fire and the competition of other plants, been reduced to a last stand in the region already mentioned. Fortunately it is rather abundant in its locality so that there is no danger of immediate extinction.

When the mats flower in February and March, they change from dull green to white and pink—pink because the buds are of that color. Later in fruit the innumerable branch tips are reddened by the color of the capsule.

These color aspects much enhance the interest in this rare species, especially when considered from the point of view of the rock garden. It has already been tried in one such garden where it is highly prized as one of the gems of the collection. However, sufficient experience has not been had with it to know whether or not permanent transplants can be made. Rare plants such as this one often give much difficulty to the grower. If it can be successfully handled, there is no doubt that this beautiful little mat form should become a favorite with all rock gardeners.

The presence of trailing arbutus (*Epigaea repens,* fig. 185) on the hot exposed deep sands of the Sandhill district is one of the surprises for the northern botanist. This species was the "mayflower" of the pilgrim fathers who loved it in their forbidding Massachusetts' forests because when it bloomed and shed its delicate fragrance, they knew spring had really come. From the shady, cool conditions of the great broad-leaved forests, where this plant is frequently seen, to the sunny, scorching white sands of the turkey-oak woodlands is a tremendous leap in habitat. The arbutus makes the leap gracefully and no one knows exactly how it does it. No other species associated with it in the forest has been able to accompany it in its excursion to the desert.

Other shrubs (and they are few in number) which are frequent in sandhill country are the following: The pinnate-leaved false indigo *(Amorpha herbacea)* with its slender, pointed clusters of blue flowers. It is more common near the coast. The sand myrtle (*Dendrium buxifolium*, fig. 85), with its aspect of a small boxwood, producing thick terminal clusters of white flowers. It is only known to occur in the Southport region. The poison ivy *(Rhus toxicodendon)* is able to grow sporadically. The dwarf locust *(Robinia nana)* is the small relative of the black locust tree but bears lovely rose-colored flowers. Its leaflet pairs are raised erect face to face in the daytime and at night are dropped until they are back to back. Near the coast is found the running oak *(Quercus pumila)*, a remarkable species with its trunk in the ground, only the branches growing up into the air. An area as great as that covered by a large house may often be observed covered with an impenetrable low thicket of this shrubby oak. Its leaves are very similar to those of the live oak.

Some Characteristic Sandhill Herbs

Variable Spurge *(Tithymalopsis Ipecacuanhae)*. Perhaps a better name would be the "long name spurge," for its technical appellation is always of interest to budding biologists, many of whom will learn it when many shorter names will be totally unremembered. It is called the variable spurge because of the very remarkable range of shape through which the leaves vary. The outline of a watermelon is strikingly different from that of an Indian cigar (Catalpa fruit), yet the diversity of outline of our spurge leaves of different plants is as great as that. In botanical words the leaves vary from broadly oblong to linear causing much distress to the newcomer botanist, who will be certain he has many species of spurge here. The plant is also notable for its long, enlarged root (Fig. 84). To set a boy to digging up the root of such an insignificant

FIG. 84. THE VARIABLE LEAVED SPURGE PUTS ITS SAVINGS INTO THE ROOT.

plant would be playing a joke on innocence. Becoming established during wet years on the driest, whitest sands, this plant with its long root thereafter maintains itself for many, many years. It is notable for being one of the pioneers on bare coarse sand sites.

WILD SANDHILL CHICKWEED (*Alsinopsis Caroliniana,* fig. 156). Without flowers this plant resembles the winter condition of our familiar moss pink. Its long stalked white star-like spring flowers make it a most valuable addition to any sunny rock garden, where it should survive any ordinary drouth.

TREAD SOFTLY (*Cnidoscolus stimulosus,* fig. 175). Contact with this humble weed-like plant will equal in experience the visitation of a small army of biting ants. The acid-loaded sharp hairs, which are distributed all over the plant, can give anyone a very unhappy half hour. A large southwestern species of this genus dots the prairies, being absolutely unmolested by cattle, so efficient are its stinging hairs.

Wire Plant (*Stipulicida setacea,* figs. 154, 155). This small, delicate annual meets its hot dry environment by carrying out a scheme of life used by many desert plants. It carries through its growth and flowering period during the spring and early summer, spending the hot months of July to September and the succeeding winter in the seed stage. So successful is this arrangement that this apparently delicate species is one of the commonest of the few plants which can grow on the bare white coarse sand.

Joint Weed (*Polygonella polygama,* fig. 152). This is a perennial of especial interest on account of its method of meeting the hot season of late summer by dropping its leaves. Curiously enough, this is just the time it blooms, so that the plants make an unusual picture with their leafless stems loaded at the branching top with the tiny white or yellowish white flowers. This behavior is so efficient in meeting the stresses of the hot, waterless sands that it has made it possible for this plant to be one of the earliest of the pioneers which can enter the glistening white sand areas.

Summer Farewell *(Kuhnistera pinnata).* This is remarkable for its late flowering period, many plants being found in bloom in November. Thus the fall sandhills are showy with the tussocks of this curious legume. Close examination of the flower will show little that is in common with pea flowers, but the pod included in the calyx shows it to be a legume after all.

Sand Myrtle *(Dendrium buxifolium).* In a few restricted areas in moister sandy soil bordering the bays, a small shrub with box-like leaves is to be found,

FIG. 85. IT IS SNOW IN THE SUMMER TIME, WHEN THE SAND MYRTLE COMES INTO FLOWER.

which when in flower presents a bright scene indeed. One of the best developments of this species of sand myrtle is to be found on the main highway between Wilmington and Southport (Fig. 85).

LUPINE *(Lupinus)*. These well known large-racemed legume plants are worthy of a much wider use in cultivation. *L. perennis,* with the many-fingered leaf, is more decorative than the *L. diffusus* (Fig. 173), which has but one, large, whitish leaflet to the leaf, making it appear most unlupine-like. The latter coarse plant, which is very characteristic of the sandhills, where its large clusters of oval gray-white leaves, arising from spreading stems, may be found in the very dry and hot scrub-oak woodlands. The lupines are difficult to transplant successfully.

QUEEN'S DELIGHT *(Stillingia sylvatica)*. This is one of the strange plants of the sandy soils. Its spreading firm branches make it appear shrub-like, yet it is completely herbaceous, dying down to the central root each winter. The branches assume a radiating aspect, very distinctive of the plant. The stamen bearing flowers are borne on a yellow thickened spike above the pistillate or seed bearing ones. After the former have flowered and furnished the necessary pollen, the whole spike falls off. Later the rather large three-lobed fruits may be observed hanging on to what is now the stem end. The leaf blades are oval with a finely wavy margin. The common name appears to have been derived from some fanciful properties possessed by the root.

MILK PEA *(Galactia)*. These are very common legumes of the drier areas in our sandhills where their trailing stems of the season, with trifoliate leaves, range for a few feet from the perennial root. The four or five species differ in leaflet shape. Examination of the calyx will disclose the fact that the two upper sepals are united, the two lateral ones are reduced, and the remaining lower one is the longest of all. The common name has no direct relation to the plant.

SANDHILL SPIDERWORT *(Cuthbertia)*. This is a delightful wild flower, with its delicate three-petaled rose-colored flowers clustered amid a mass of erect grass-like leaves. If the leaves are less than one-eighth inch across the plant is *C. graminea;* if larger, it is *C. rosea.* These plants grow successfully in the most sterile and dry sands, where they add to the restricted flora found in such places. They deserve a wide introduction to sunny rock gardens.

SANDHILL MOSS PINK *(Phlox Hentzii)*. This species differs from the common moss pink *(P. subulata)* in that the plants do not form large, loose mats, but the stems stand stiffly erect in small masses. When in flower the stem cluster is hidden by the bloom. It is one of the most desirable plants for exposed rock gardens, being much superior to the other species. *P. Hentzii* is distinctively southern and is confined to sandhill country, while *P. subulata* is widely spread in the north as well.

SANDHILL TREES AND SANDHILL SOIL

The sandhill vegetation illustrates in a most vivid manner the control of plant distribution locally through changes in soil water conditions. Some of the above mentioned shrubs, for instance, are to be found in shallower sand with clay subsoil, but the surest way to tell the subsoil conditions is by use of the trees. In the great sandhills, every illiterate countryman knows that black jack oak *(Quercus Marylandica)* will tell you where the clay layers occur at or near the hill tops. More or less frequently seen with this familiar oak are the blue-jack oak *(Q. brevifolia)* and the dwarf post oak *(Q. Margaretta)*. The former is a very attractive tree with its bluish-white elliptical leaves. Along the slope bases, where the water conditions are still more favorable, the white hickory may be found. To one interested in plant distribution it is an interesting experience to note the tree zones as one leaves the turkey oak and pine hilltop and walks down the slope. The sharpness of the limits of the hickory and the oaks will be surprising.

One valuable lesson to be learned is that the dry

soil plants continue down among the moisture preferring kinds. That is, such plants do not refuse a drink when they can get it. But those whose nature demands better soil water conditions are forever prevented from surviving in the drier places.

An even more pronounced case of soil control in distribution than that seen on a single coarse sand slope with hidden clay layers, is that of the red hills. Scattered here and there in the sandhills are high points where the soil changes to a deep rust color due to the high amount of the cementing iron oxide in it. At places of rapid erosion the substratum is observed to be sandstone, which accounts for the successful resistance of such hills to weathering forces. Blue's Mountain, in the Fort Bragg reservation, one of the highest points in the region is a "red hill." Of preeminate interest, however, is the fact that on these hilltops may be found, if not cut over, excellent examples of the great broad-leaved moisture-loving forest so characteristic of the piedmont country. So retentive of moisture is this red sandy soil and so greatly improved is the capillary function, that the white oaks and other familiar representatives of its social group march right over these hill tops. Following fire and cutting, the loblolly pine comes in just as it does in the piedmont. In the old fields it forms the familiar dense pine thickets. And even more striking, in the recently abandoned fields on such hilltops, the familiar piedmont broom-sedge is seen in excellent stand.

The Story of the Sandhill Soil

Now that our reader has been introduced to some of the contrasts in our sandhill plant societies, he may wish to delve a little further into the soil science underlying these differences.

All soil stories begin with geological tales. Always must we look to geology for the primary explanation of our vegetational contrasts due to soil conditions. Things that happened many millions of years ago are today directly responsible for this type of forest or that, for this failure in peach growing or that success with dewberries. The marvelous unity of nature—its interrelatedness in both space and time—is a source of constant wonder to the scientific mind. No better illustration is to be had of all this than that brought out in the story of the sandhills.

Many, many millions of years ago during a period the geologist refers to as the Lower Cretaceous, the sea was well in over the eastern one-third of the state. Rivers from the piedmont flowed into it so that many and diverse conditions for depositing materials were created. During this time alternating layers of sand of widely varying composition were laid down. Then the land emerged from the sea slowly, and for a few million years was exposed to the wearing down action of running water, becoming changed from its original nearly level condition into hills and valleys. If one could have walked about on that ancient land, he would have noted the white clay-like fine sand layers and the various colored coarser sand strata, but nowhere would he have encountered a uniform top layer of medium to coarse sand. This latter deep loose sand was still to be put down by the sea after a new depression of the land had taken place in the very recent geological time of a million or so years ago. So different is this mantle of sand spread over the land by this, the last advance of the sea, that the geologist has given it a special name "The Lafayette layer." The very much older layers below he calls Patuxent of the Lower Cretaceous. And thus it is that today our pure turkey-oak and long leaf pine lands in the sandhill district are confined to the deep loose "brown sugar" sand of the Lafayette layer. But where the Patuxent clays get near the surface, they call for the black jack and other oaks, and on the red sand hills, originally laid down in the same dim, distant Cretaceous, we have the luxuriant white oak-hickory forest. And so it is that the influences in nature extend not only through vast spaces but through vast time as well.

We must not leave the geological story without mentioning a minor change of some interest which is still going on. Nothing is so out of favor in the coastal plain as a dust storm. In dry weather of high winds, the dust from the finer soils of regions surrounding the sandhills is carried over the coarse sands and deposited on the surface. Through the recent thousands of years the accumulation has been sufficient to be readily detectable. One must examine, however, the soil on the flat hilltops where the surface two to five inches will be found to run finer than the layers below. On the slopes the finer material is washed away by the drenching rains leaving only the coarser sand. The presence of the finer material in the surface layer tends to improve the soil in a number of ways. It is perhaps for this reason that the wire-grass uses this shallow layer as a habitat for its roots.

The relations of soil to vegetation should be known to land buyers and farmers. Such information will prevent persistent losses in agriculture. Avoid land where no black oak *(Quercus velutina)* occurs. If white oak and dogwood are there, let go of the money; the land will be the best the sandhills can give one.

Why a Desert in a Land of Rainfall?

The coarse sand soils and their vegetation have a desert aspect. Why should this be true when they lie in a region of abundant rainfall (40-50 inches)? Only the more important relations may be pointed out here. They should be of interest to all plant growers, for they embody certain basic principles of soil science, most valuable to know.

Strange as it may seem, the amount of surface on the particles in one gram of sand is but one-eighth of that on the vastly greater number of particles in one gram of clay soil. This fact has its greatest significance in relation to retention of the valuable mineral nutrients. So leached and low in these nutrients is the soil from a turkey-oak and long leaf pine woodland that tomato plants grown in it, which were watered daily for six weeks, attained the height of one inch and never developed secondary leaves. Of especial interest, however, on the other side of the question, is the discovery that when nutrition is furnished as in soil fertilization, the plants are not only benefited directly, but also indirectly during drouth periods through the stimulation of root growth. Since in coarse sand water does not move readily by capillary action, if the plants are to get the water still available, they must "grow for it" so to speak. Without adequate nutrient they cannot do this; hence in a state of nature only those plants can survive in deep coarse sand which check the

FIG. 86. THE USUAL UPLAND FIRE-OPENED LAND. CONTRAST WITH NEXT PICTURE TAKEN BUT A QUARTER OF A MILE DISTANT.

outgo of water above ground, and therefore need very little of that most important life giving substance.

Thus an added reason for maintaining high fertility in sandy soils has been found. Well nourished plants are certain to come through long drouths vastly better than unfertilized ones.

THE SANDHILL VILLAIN

Having taken up the plants of our sandhill society and their soil homelands, we must add a brief discussion at least of the villain, or ogre, who periodically makes his visitations in all of the sandy regions. The losses from his depredations run into millions. He has been coming to the sandhill country all too frequently for thousands of years. The Indians knew him well, though they feared him greatly. Strangely enough,

FIG. 87. A BLACK JACK OAK AREA SURROUNDED BY FIELDS GIVING FIRE PROTECTION IN RECENT TIME.

his visits have been increased since the white man's advent. Government aided by science has not conquered him. Like the ruthless gangster he defies society successfully. He is seldom caught, for his allies are too strong. His name is Fire.

The coastal plain in general and the sandhills in particular suffer more from conflagrations than the piedmont and mountain regions. The reason for this is a very simple one. The sandy soils, both bog and upland types, carry plant structures low in water content—inflammable vegetation. Fires will occur in such regions despite every control measure. Even lightning may, and often does, start them.

The vicious circle of the three elements, sand, plant tinder, and fire is almost impossible to break. Ecological theory holds that soils are constantly improving in nutrient and water retaining capacity. The accumulation of leaf litter alone would do much in those directions. With an improving soil the plants of higher water content, and therefore of higher fire resistance, would tend to come in and our sandy areas would be eventually forested by a tall broad-leaved forest. But the soil building process is entirely too slow; the inevitable fire turns the valuable litter into smoke, clears away the wire-grass tussocks, and sets back the shade making scrub-oaks. When all this has happened, the habitat is hotter and drier than ever so that only inflammable plants may enter, and they encourage fire. And so the vicious trio of sand, vegetational tinder, and fire will, for many years to come, hold their reign in spite of all that man can do about it.

Some compensation lies in the fact that the curious sandhill vegetation will thus be present for future plant lovers, who will spend many very profitable hours delving into the mysteries of adaptation, and others who will come for a few beautiful flowers and some interesting foliage with which to decorate their sunny rock gardens.

CHAPTER VIII

THE MELTING POT: OLD FIELDS WHERE PLANT FOREIGNERS AND NATIVES MINGLE

THE REGENERATIVE powers of organisms constitute one of life's most important aspects; only living things can heal their wounds. The broken or maladjusted machine is utterly helpless and completely dependent on outside forces for its repair. But not so is life.

It is a most curious and interesting fact that this restorative activity of plant life extends far beyond maintaining the individual and its parts into the realm of plants in the mass. By this statement we mean that if the original society of plants in an area is destroyed —a wound made in the primeval forest cover—nature promptly proceeds to heal that wound, and in time the affected area will be just as it was. Though a scar is left after a gashed finger, the healed forest will show no evidence of the injury it suffered, so perfect will be the work of restoration. Naturally, years, instead of days, are involved, but immediately following the injury or destruction of any original native vegetation, the advance guard of certain species begins to reclaim the bare area and initiates a series of plant communities which follow one another until at last the original plants reënter and the green epidermis of the earth is fully restored.

This fascinating story of plant mass regeneration which goes on everywhere where plants grow and are destroyed, is familiarly known as plant succession, and it is the early chapters of this story with which we are concerned in our study of old field weeds. Once we realize this fact clearly, it will add much to our interest in abandoned lands, and instead of thinking of the old field plants as useless weeds we shall recognize them as most valuable pioneers in nature's work of rebuilding her devastated forest. As it will be shown later, without their coming the rich woods could never reappear.

For the plant loving gardener these facts will ameliorate his or her bitterness against the ever persistent weeds which perennially war against the much loved things of the flower garden. These "terrible" weeds are merely shock troops in nature's determined attempt to reclaim her territory. While they are exasperating, we may all agree that it would be more terrible not to have weeds quickly reclaiming our unused lands, where they not only prevent these lands from being cut into gullies, but, of vastly more importance, prepare them for the return of the forest. Before we have finished this chapter we shall hope to have every reader as ardent as the author is, in the defense of weeds. Unlike mosquitoes, weeds are not wholly bad.

How the Plants Reclaim the Old Field

Tenant-farmer Smith leaves his lands for the mill town in piedmont North Carolina and no one is found to take his place. His cornfield at the base of the hill where the soil was moist and mellow was choked with crab-grass even before the summer was over. The field

FIG. 88. IT IS FALL WHEN THE BROOM SEDGE FLOWERS AND FRUITS.

could scarcely be called abandoned, for the crop stalks of the season are still standing very erect all over it, and yet the crab-grass has made an almost perfect cover on it, as if an unseen hand had sown a second crop.

The following year no plow again disturbs the soil and the land lies open to all plant comers, but only a few constitute the second army of invasion and, of these few, some one kind will be apt to constitute the innumerable privates in the ranks. These may be ragweeds, which will come up thick and tall.

Another summer comes, and in the abandoned cornfield, becoming noticeable in the fall, are some yellow-brown stems with tiny white hairy fruits, standing here and there among the ragweeds. The third wave of invasion is on, and during the next year, the third after the abandonment of the field, the ochre-colored

FIG. 89. A HEAVY "CROP" OF PURE BROOM SEDGE INDICATES LITTLE COMPETITION FROM PINES DUE TO THE ABSENCE OF THE LATTER FROM THE VICINITY.

broom-sedges (Fig. 88) will have made their impress in the story of regenerating Farmer Smith's field. Once present the broom-sedges spread rapidly in moist soil, and by the end of the fifth year broom-sedge tussocks will be thick over the area with few weeds of any other kind left to compete with it (Fig. 89).

Peeping above the surface of the broom-sedge a year or two later will be the green tassels of the loblolly pine, which will here almost certainly come up in a thick stand, and when they get above the broom-sedge, the latter will thin out rapidly. The pines will go on growing for many years in complete charge of the one-time cornfield, but they in time must go the way of their predecessors, for the oak-hickory forest will come right up in its midst, shading and killing the lower branches first and finally catching up with the smaller trees, which die out in large numbers. It may be fifty or eighty years more, but what are decades in the lives of plant communities which are thousands of years old? The last chapter of this history of invasion is not yet written, for on this cool north-facing hill base, the beeches and maples will, in two or three or more centuries, gradually replace the oaks and hickories, and, once this type of rich woods has captured the old field, it will, as the trees die, replace itself through coming millenniums. Nature has reclaimed Farmer Smith's field and put it back in the same condition that the Indian saw it when he passed that way on his deer hunts, three hundred years ago.

If the reader has that kind of mind which makes him curious as to why things are as they are, he will immediately want to know why nature must go through all these stages and sacrifice, much as armies do, one wave of troops after another before stability and victory is finally won. "Why," one can hear him say, "can't the beeches and maples come directly into the cornfield, or at least the oaks and hickories? Surely the pines ought to do it." But they don't. And who ever saw broom-sedge capturing an abandoned field the second year?

Let us preface our answer by stating at once that all of the factors which are operating in carrying on these changes are not yet known to science, at least as to their relative importance. But certain suggestions may be made which help much in our understanding of this interesting story of plant succession in our old fields.

FIG. 90. THE TALL BEARD GRASS IN OLD FIELDS TELLS THE LOCATION OF THE MOISTER SOILS.

VARIATIONS FROM THE TYPE SUCCESSION

Crabgrass *(Syntherisma sanguinale)*, together with Bermuda grass *(Cynodon dactylon)*, which resembles it, is the remarkably uniform pioneer all over the state except on the deep coarse sandy soils of the sandhill districts in the southeastern quarter of North Carolina. Here the horseweed *(Leptilon canadense)* or a new weed which has recently come in from the south, the yellow aster *(Isopappus divaricatus)*, is prominent as the first comer. They are both tall weeds and so we may say that the low cover stage of crab-grass in the sandhill uplands is left out. Where the coarse sand forms a thin cover over water-holding clay layers below and in the valley bottoms, the crab-grass will be seen as usual making its attack on the bare area.

Where the sand is medium to fine and on all of the loam and clay soils not too dry over the entire state from near sea level to the highest fields in the mountains, one may expect crab-grass to be patiently awaiting its chance to steal the field and start the area on its way to forest again.

The seeds of crab-grass (and it seeds profusely) may lie dormant in the ground for many years, and it is due to this fact that it can, with almost magical swiftness, appear when its chance comes. A few coarse weeds like horse-nettle *(Solanum Carolinense)* and cocklebur *(Xanthium canadense)* will almost always be found coming in with crab-grass but in very open stand.

The second, or tall-weed stage as we should call it in old fields not excessively dry, is remarkable in the wide variation of weeds which may be locally dominant. One may readily note this fact when driving through the country. Here will be a field covered by a thick growth of heath aster *(Aster ericoides),* while across the road will be a field made just as attractive by the tall plume-like masses of fine foliage of the southern dog fennell (*Eupatorium capillifolium,* fig. 91). A half a mile farther, and horse weed or ragweed is in full control. If cattle are allowed to graze, the bitter weed *(Helenium tenuifolium)* will have its chance even though the cows may eat enough of it to flavor the milk. Golden-rod must not be overlooked, a number of species being represented. Of especial interest is the eastern red-stemmed golden-rod *(Solidago fistulosa),* which is sure to be found on muck or peat lands where it may struggle for temporary control with a tall panic grass *(P. verrucosum).* On this latter soil dewberry

FIG. 91. SOUTHERN DOG FENNEL LUXURIATES IN THE OLD FIELDS OF THE LOW EASTERN COUNTRY, IF THEY ARE NOT TOO WET. WHERE THIS WEED GROWS THE DRAINAGE WILL BE GOOD.

also is frequent covering extensive areas (Fig. 92). These tall weeds bring death to the crab-grass by cutting out the light.

In the upper piedmont and mountains such showier weeds as the following will be more prominent: wild yellow lily *(Lilium canadense)*, wild carrot *(Daucus carota)*, evening primrose *(Oenothera biennis* and *O. laciniata)*, many golden-rods and asters, and the field daisy (*Chrysanthemum leucanthemum*, fig. 209), a plant said to be the state flower of North Carolina but this cannot be true because this weed is an immigrant from Europe (Fig. 93). A state with a flora as rich as that of North Carolina does not need to elevate a plant foreigner to such a high place. Among other mountain weeds of the second stage which deserve mention

FIG. 92. THE WILD DEWBERRY OCCASIONALLY CATCHES AND HOLDS AN OLD FIELD ON MUCK SOIL.

are flowering spurge *(Euphorbia corollata)*, mountain mint *(Koellia pycnanthemoides)*, blackeyed susan *(Rudbeckia hirta)*, Joe-Pye-weed *(Eupatorium purpureum)*, fleabanes *(Erigeron)*, ironweeds *(Vernonia)*, and flat-topped golden rod *(Euthamia graminifolia)*. In the moist meadows of the piedmont district no more beautiful plant grows than the paint brush *(Castilleja coccinea)* with its scarlet tips (Fig. 94).

THE PIONEERS ON DRIER FIELDS AND ROADSIDES

All of the foregoing discussion refers to old field vegetation where the water supply is fair to abundant. On hilltop fields and especially those of a stony or gravelly nature where short drouths bring the water to a low point, the weeds are very different; yet there will be an inevitable succession. Here the well named pov-

FIG. 93. THE RIOTOUS OLD FIELD OF THE MOUNTAINS; THE TALL WEED STAGE IS MADE UP OF A WIDE RANGE OF COLORFUL SPECIES.

erty grasses (*Aristida oligantha* and related species) will have their chance. The other weeds, even including crab-grass, will find it too dry, and the slender poverty grasses and drouth adapted weeds will gain their place in the sun without a contest.

So similar are roadsides to dry sterile fields in the nature of the soil habitat, that they can generally be treated in the same discussion, for many of the weeds are common to both. Among the most familiar are the bracted plantain *(Plantago aristata)*, the creeping button weed *(Diodia teres)*, the pinweed *(Lechea Leggettii)*, the naked beard grass *(Gymnapogon ambiguus)*, pepper grass *(Lepidium Virginicum)*, bitter weed *(Helenium tenuifolium)*, needle leaved panic grass *(Panicum aciculare)*, lance-leaved plantain *(Plantago*

FIG. 94. THE PAINT BRUSH OF THE PIEDMONT WET MEADOWS.

lanceolata), and undersized low ragweed *(Ambrosia artemisiifolia)*.

Occasionally fields will be taken by small weeds that ordinarily are not widespread. Among such are the wild onion *(Allium vineale)*, sheep sorrel *(Rumex acetosella* or *R. hastatulus)*, bracted and lance-leaved plantain or buckhorn *(Plantago aristata* and *P. lanceolata)*. In the early spring, or even showing up in the winter when the warm south wind blows for a number of days, will be the chickweed *(Stellaria media)*. A real old field curiosity is one in which the ground is covered thick like a deep soft carpet by the German knot grass or knawel *(Scleranthus annuus)*. This weed resembles chickweed somewhat but has stiff awl-shaped leaves and small green flowers in bunches at the branch ends. This weed has unquestionably come in through adulterated seed. The winter cress *(Barbarea barbarea)* and wall cress *(Stenophragma Thaliana)* are unimportant though seen in large numbers in the spring. In the eastern section, nut grass *(Cyperus rotundus)*, which is not a grass at all but a sedge having tuber bearing underground stems much like an Irish potato, may give a deal of trouble to the truck gardener and others. Fortunately this bad weed prefers moist soil and does not spread to well drained areas. Occasionally a plant commonly thought harmless will flare up as a pest. One striking case of this sort was that of the blue bottle *(Muscari botryoides)*, which gave some farmers in the piedmont region no end of trouble.

A very harmless and attractive weed is the blue toadflax, which makes many fields cerulean in the early spring. It is a winter annual in the South and the thousands of slender flower-bearing stems arising from the basal rosettes give a most distinctive spring aspect to the piedmont lands.

As one drives along the hard surfaced roads in summer it is surprising to note how variegated the weed flora is. For a quarter of a mile some one weed will be dominant; then perhaps will come a mixture of many species followed by a brief pure stand of something else. It will shorten many a long drive to watch the weed flora on the road shoulders where they are doing their best to beautify the raw earth and, of more practical value still, hold the soil together, making the task of the highway foreman easier.

Broom-Sedge

On both the moist and dry fields all these varied species of the second stage tend to give way to one genus of grasses; viz., the omnipresent broom-sedges *(Andropogon).* The broom-sedges (not sedges at all) are a very distinct group of grasses of which there are about six species in North Carolina. Of these some are much more common than others and of the latter are types adapted to the wet, the moist, and the dry soils, where they all play their rôle in taking over the fields from the tall or middle stage weeds. In the wet or muck lands it is *Andropogon corymbosus,* in the moist fields it is *A. virginicus* or *A. Elliotii,* and on the dry areas *A. argyraeus,* the silver beard, looks out for the succession. They are all tussock grasses which after attaining dominance give a most characteristic aspect to the landscape. The North Carolinian who goes to New York State, for instance, will soon miss the broom-sedge; a most colorful note of our landscape both in summer and winter will be lost.

The broom-sedges have the unusual habit of delaying the growth of the fruiting stalks until fall, when they quickly shoot up and set their hairy wind borne tiny seeds free on the cool October and November breezes. These stalks then remain standing for a year until the next ones come on to take their places, so that the broom-sedge field seems to always look the same. The broom-sedge is a perennial and by stooling out at the sides gradually enlarges its tussock. As the tussock grows in age the number of fruiting stalks increases so that one with practice can tell approximately how long the broom-sedge has been in the field by noting the stalks on the largest tussocks. The broom-sedges apparently conquer the tall weeds through root competition.

To the writer broom-sedge fields against the pines

are very picturesque, fit subjects for the work of a landscape painter. In addition, however, to its aesthetic contribution, broom-sedge makes an intensely practical one, a gift to North Carolina and the South of the greatest importance. We here refer to the fact that were it not for the broom-sedges much of piedmont North Carolina would resemble the bad lands of the Dakotas, where erosion unchecked has cut the land into innumerable gullies making it totally unfit for any use whatever. Due to erosion of the cultivated lands, it is estimated that seven inches of soil goes off the hills of the piedmont in one hundred years. But when the farm goes back to nature, in just a few years the sure grip of millions of broom-sedge roots is felt in the soil and little is the loss of earth from an area with a good stand of the yellow broom-sedge on it. In its silent way broom-sedge is worth millions of dollars to North Carolina folk. So let me suggest to my reader in the piedmont that the next time you pass a broom-sedge field, pause a moment and in imagination picture the valueless areas of bare, red clay ridges running helter-skelter in all directions which would surely be there were it not for the early capture and preservation of the land by this grass, which holds it well and, in turn, eventually passes it over to the even better safe keeping of the forest.

And while we are presenting the practical aspects of this grass, mention must certainly be made of the value expressed in its common name. What a sweeping the floors of many a cabin and even more pretentious homes have had with the bundles of it! Dollars saved by the use of this gift of Nature to domestic economy, would run into thousands.

FIG. 95. BORDER OF SEASIDE PASPALUM GRASS *(Paspalum distichum)* ALONG CURRITUCK SOUND WHICH OCCURS IN ZONE OF WINTER HIGH TIDES. SUMMER TIDAL ZONE NEXT TO IT AT RIGHT. OLD FIELD WEEDS AT LEFT.

A MINOR WEED PROBLEM

A minor but interesting situation in weed relations is to be observed along the sound shores. In Figure 95 is shown a bit of the main land border of Currituck Sound. The low grass zone between the bare tidal area and the tall weeds is a pure stand of the seaside paspalum grass *(Paspalum distichum)*. Why this definite zone? Apparently this perennial grass tolerates the salt water of the high winter tides, which kills the other weeds. The salt water thus acts as a selective factor producing this very definite border over long stretches of the sound shore.

Such minor problems in distribution as this one are continually being encountered by the ecologist, who immediately wishes to know the environmental factor complex which will solve it.

Garden Weeds and Their Control

Because of the higher fertility of gardens, together with the fact that many of the flower gardens during drouths get rain that has been pumped through city mains, weeds appear with a vigor and a vengeance not seen in their advent into the fields. Most of those previously mentioned may be expected, but especially to be added are the pigweeds especially the red root and spiny amaranth *(Amaranthus retroflexus* and *spinosus)*.

Weeds in gardens compete with the valuable plants for water and nutrients to a far greater degree than most folk imagine. During a drouth period the presence or absence of weeds may determine the survival or death of the garden species. They are in every sense enemies to the death, of the valued plants. When the numbers of seeds which can be formed by single weeds is considered, it becomes apparent why the gardener has such a fight on his hands. One wild carrot plant can set twenty thousand seeds. Shepherd's purse, a minor pest, may produce sixty-four thousand seeds. A large pigweed is said to drop some one hundred thousand progeny. The numbers are always high.

In a work of this character space limitations will only permit a most general treatment of this important subject. In fighting particular pests one must know their nature—are the weeds annuals, biennials, or perennials? Only the most general rules, however, may be given here. Above everything, weeds should be prevented from going to seed. The seeds of the pests should never be brought to the farm or garden in adulterated seed or otherwise. This is the ounce of prevention. For the established perennial, frequent cutting back, preventing top growth, will starve out the underground stems. Persistent cultivation is recognized by all as the great weed deterrent. For the flower garden nothing has yet been invented that is more successful than the human hand weed puller when the soil is mellow after rains. Chemical sprays are used only in special cases of mass invasion. Sodium chlorate (one pound to a gallon of water) was found at the Michigan Experiment Station to be very successful against quack grass. Special precautions must, however, be taken when any of these chemical sprays are used. The use of heavy tarred paper of various sorts to cover the ground around the garden plants is highly successful since the weed seedlings can receive none of the all important light. Aesthetical considerations will, however, preclude the use of such ground cover in the case of flower gardens. Imagine roses or iris arising from the overlapping sheets of dull black creosoted paper! This method may be excellent for the commercial trucker, but hand weeding is destined to go on for a long time in the grounds of the flower lovers.

For the encouragement of North Carolina plant growers, it might be pointed out that we do not have within the state such terrible weed invasions as those of the Russian thistle and the western field bindweed. These pests have literally swamped farms and forced the land-owners off from their acres just as mercilessly as if they were bandits with guns. Many a rural tragedy in the West can be wholly traceable to these vicious gangsters of the plant world.

CHAPTER IX

THE GREAT FOREST: THE UPLAND SHADE GARDENS

JUDGED by every standard this is the greatest of all our plant communities. In size of plants, in number of species, in extent of area, in diversity of structure within the community, in its contribution to gardening, the great upland forest of both broad-leaved and coniferous trees, with its host of subordinate plants, constitutes our richest vegetational heritage.

As the reader has already discovered from all that has gone before, water is the principal key to the understanding of the delimitation of the communities. With regard to this highly important factor, the great forest occupies a middle position. This vegetation prefers the soils which are neither excessively wet for long periods nor become excessively dry or low in water supplying power. Such conditions of good drainage coupled with sufficiently high retentiveness of the soil water may be found in extensive or restricted local areas in all parts of the state. The most typical regions exhibiting this community in second growth condition, however, are the central piedmont and the mountain slopes under five thousand feet. The medium to fine sands of the coastal plain carry one phase of it, and since these soils are much more abundant in the northern half of the coastal plain, we must add that region so far as uplands go, with of course many local areas in the southern half.

The North Carolinian of today who speeds across the open country in his fast car, can hardly imagine what his state was like when the original forest covered it. At the time Columbus was becoming acquainted with America's shore line, the plants were in almost complete domination of this country. The Indians were not present in sufficient numbers, nor did they have adequate tools to interfere seriously with the processes going on in the leafy kingdom. One exception must be noted in passing. They did fire the woods occasionally. But in spite of this, over vast areas stretched the unbroken forest. On the trails through it, one could travel for days without a good view of the sun, and at night the constellations could seldom be seen because of the interfering canopy of high verdure.

The streams, thanks to the omnipresent forest and its spongelike floor, were of clear water most of the time, and their flow did not fluctuate widely as at the present. Animal life was abundant in all regions, the larger species ranging throughout the state. In its pristine condition the entire eastern United States was a naturalist's paradise. And within this vast virgin wilderness, the maximum attainment in size and kinds of trees was in the southern Appalachian mountains.

The great forest may be classified into a number of sub-communities which are still related to habitat—whether wetter or drier. On the dry side we have such a forest type as that made up of black jack oak, post oak, yellow and loblolly pine, whether observed singly

FIG. 96. THE BEECH IS AN INDICATOR OF GOOD SOIL. BEECH WOOD SOIL IS EXCELLENT FOR GARDENS.

or collectively. It is this kind of forest which grades into the extremely dry forest of long leaf pine and turkey oak of the sandhills, which community should not be included in the great forest because of its distinctive character. On the wet side we have such aggregations as the beech (Fig. 96), maple, and sweet gum, taken singly or collectively, which, in turn, lead over into the very different swamp forest associations. In between come the extensive complexes of oaks and hickories, so characteristic of the piedmont region, and the oaks and chestnut, so characteristic of the mountains except on the summits.

Another very characteristic subdivision is that which is found on the flood-plains of rivers, a community made up in part of river birch, sycamore (Fig.

FIG. 97. THE SYCAMORE IS WHOLLY CONFINED TO RIVER BOTTOMS EXCEPT WHEN TRANSPLANTED. VIEW UNDER AN OLD BRIDGE ACROSS THE NEUSE.

97), hackberry, blue-beech, swamp white oak, and red maple as principal species.

So remarkably delicate is the adjustment of vegetation to soil conditions, particularly the water factor, that on a single hill of the piedmont or upper coastal plains may be seen excellent examples of all three forest types, the dry, the medium, and the moist. Near Raleigh is an extensive forest area bordering a branch of Crabtree Creek on the site of some formerly worked graphite deposits called the Lead Mines. On the level hilltop is a woodland of black jack oak and some post oak, at the beginning of the slope the yellow pine is frequent. On the lower two-thirds of the slope are excellent second growth stands of white oak and mockernut hickory and numerous other medium water trees

FIG. 98. THE FLOWERS OF THE MOST BEAUTIFUL OF OUR FLOWERING VINES, THE YELLOW JASMINE, ARE FREQUENTLY SEEN IN PINE WOODS.

and shrubs. At the base near the spring is a practically pure stand of beech surrounded by mixed beech and maples. These latter trees are rooted in very moist soil, kept in that condition by the underground seepage down the slope.

In these different forest types the shrubs and herbs are different also. In gathering native material for gardens one should always keep in mind the type of forest from which it was obtained. The writer has seen the rich, moist, forest hepatica and wild ginger placed in open sunny rockeries where they are doomed to death in the first season after transplanting. And it would be just as great an error to move the attractive blazing stars and white-topped asters of our dry woodlands into heavy shade.

The Problem of the Pines

Of unusual interest in forest distribution is the problem of the pines, especially the loblolly pine which has enlarged its range greatly in the last seventy-five years. Originally it was dominant only in the northeastern corner of the state. At the present, following the practical annihilation of the long leaf pine forests and the great reduction of the deciduous woody vegetations of both coastal plain and piedmont, the loblolly pine has largely taken over these areas except the extremely dry coarse sand ones. With broom-sedge it is common on many savannah-like areas in the coastal plain.

The pines may always be considered dry forest types. But dry forest conditions may be readily created in moist forest areas by the simple expedient of destroying the deciduous trees by cutting or burning. This change makes possible the rapid entry of the shade intolerant pines. Thus it is that wherever one observes a thick close stand of pines of about the same height, one may be very certain that a field was abandoned, for in an old field there are no oaks or other trees to sprout immediately and give the young pines serious competition. These groves of pines are especially prominent in the piedmont so that if one wishes to observe how much land has gone out of cultivation, as he tours the piedmont roads he has but to note the pine groves of all ages and sizes.

In the piedmont region these pine woodlands (Fig. 99) are all destined to disappear under the silent slow attack of the oak-hickory forest which inevitably will come in because of the favoring clay soil of high water supplying power. As the shade producing trees de-

FIG. 99. OAKS AND HICKORIES WILL CONQUER THESE LOBLOLLY PINES IF FIRE DOES NOT INTERFERE.

velop amid the pines, the lower branches of the latter will go first, followed by the higher ones, until finally only a few of the tallest trees will survive, holding their pine-needle tops above the luxuriant shade-making foliage of the maturing broad-leaved trees. To transform these pure pine stands to a rich oak-hickory forest may take a hundred to five hundred years, depending on whether the site is moist or dry, but in the economy of biological nature, this plant succession, as it is called, is one of the certain movements.

On the medium to fine sands of the coastal plain, the same process is taking place but in the sandy soil region, the lower water supplying power combined with greater soil sterility due to leaching, so checks the growth of the broad-leaved forest that before it can become established, even in an immature stage, the inevitable fires will devastate it. Young pines, beyond the seedling stage, will survive light ground fires and the older pines will come through heavier fires unscathed. Hence it may be easily understood that the occurrence in the coastal plain of such extensive pine areas is in part the result of fire. These pine forests are then spoken of as fire subclimaxes.

Many ecological authorities in this country hold to the conception that, given time enough (two to three thousand years) for the soil moisture and nutrients to be built up to a high degree, the sandiest coastal plain soils without fire sweeping through their vegetation, would eventually carry the climax oak-hickory forest. Seen from a long-time viewpoint, some fire is an essential to the production of the pure pine forests. On the other hand, an excess of fire both as to frequency and intensity may be destructive to these same forests. The southern foresters are involved in a perennial debate upon this interesting question of how much or how little fire it takes to develop maximum stands of long leaf and other pines. There can be little doubt, however, that the coastal plain pine forests are true fire sub-climaxes.

Forest Types

The medium-watered forests in the coastal plain are found well developed only on the margins of swamps or on the level flood plains of the larger rivers. On the swamp margins white oak, southern red oak, water oak, willow oak, and a little hickory are to be expected (Fig. 100). In the moister portions, beech, tulip poplar, and sweet gum are not uncommon, especially the latter is always to be expected in moist but not permanently wet soil.

FIG. 100. WHITE OAK FOREST LAND IN THE COASTAL PLAIN IS SUPERIOR FOR AGRICULTURAL PURPOSES.

On the piedmont Pinchot and Ashe recognize three forest belts, an eastern and western pine region with deciduous forest between. In accordance with expectations the pine forests are correlated with coarse grained soils or thin surface ones introducing the dry and sterile conditions always to be correlated with the needle-leaved trees. In the eastern region the loblolly occurs with the yellow pine, while in the western zone the scrub pine is associated with it. Many oaks and other broad-leaved trees are naturally associated with the pines. In the western area the scarlet oak for some reason is especially prominent, while in the eastern zone the black, post, and black jack are the principal broad-leaved trees. The great central piedmont region of clayey soils is low in pines except those which are successional. The original forest was almost free from pine timber. Here the white, black, southern red, and scarlet oaks, together with the hickories, were prominent.

The lower mountains or those which one first meets upon approaching the highland district from the piedmont, carry a forest of somewhat different composition than the high mountains beyond. These lower mountains are carved in the eastern edge of the Asheville plateau and may be roughly included between the altitudes of one thousand five hundred and three thousand feet (Fig. 101). On these slopes formerly grew magnificent forests of white, chestnut, black, northern red, and shingle oak, together with a wide array of other large trees such as the butternut, shell-bark hickory, and chestnut (Fig. 102). At the higher altitudes in these mountains excellent stands of the very beautiful white pine occurred.

In the higher mountains above three thousand feet and ranging up to five thousand and in places to five thousand five hundred feet, we find the great oak-chestnut forest. The chestnut, formerly the dominant tree, is now disappearing under the insidious attack of a fungous parasite, but its place will be rapidly taken by the oaks and other trees. The virtual elimination of a dominant tree species such as the chestnut is a major event in the plant world and the problem of the adjustment of the other trees in filling in the gaps made by the dead chestnuts is a matter of the greatest interest to foresters. In any event, in the near future the ecologist and forester will speak of the "higher mountain oak forest."

The commoner oaks are the white, northern red, chestnut, and black. In addition to the oaks, and espe-

FIG. 101. NATURE RAPIDLY RECLOTHES HER MOUNTAIN SLOPES; SECOND GROWTH TIMBER NEAR CHIMNEY ROCK.

FIG. 102. A SILVER THREAD WINDS THROUGH THE GREEN PLUSH OF THE FOREST; A GLIMPSE OF THE ROAD IN A VALLEY BASE NEAR CHIMNEY ROCK. THE TREES ARE CHIEFLY SECOND GROWTH CHESTNUT-OAK.

cially abundant on the cool north slopes and in coves, are the hemlock (Fig. 103), cucumber tree, tulip poplar, beech, and sugar maple. At the higher altitudes, the buckeye, sweet and yellow birch, with beech and chestnut in dwarfed form, occur. The great rhododendron *(R. maximum)* is also common (Fig. 104).

Any visitor to the higher mountain slopes will be much interested in noting how distinctive the upper zone of deciduous trees is in stature and composition. The chestnuts, if they are still surviving the blight, become lower until scrub trees only twenty feet or less high may be observed. The dwarf beech groves are especially interesting and extremely picturesque. An excellent example of a dwarf beech forest may be observed from the horse trail which goes up the state line ridge of Mt. Collins from Newfound Gap in the Smoky Mountain National Park. Its presence here is especially interesting since it occupies the ridge top above the balsam-fir forest on the steep slopes below. Occasionally pure forests of repressed northern red oak are encountered, the trees not ranging above thirty feet in height (Fig. 105).

The mountain forest complex just described is to be seen at its best in the Smoky Mountain National Park just mentioned in a minor connection. On clear days from all of the high peaks, LeConte, Guyot, Collins, Clingman's Dome, the most amazing masses of verdure are spread below the observer on the precipitous slopes, a green verdure which contrasts sharply with the deep, almost black-green foliage of the balsam-fir forest of the higher and cooler upper slopes. Those who have not yet visited this newly created park can have little idea what a glorious spectacle awaits them

FIG. 103. THE FEATHERY HEMLOCKS LIFT THEIR CROWNS INTO THE SKY.

FIG. 104. WAITING FOR INSECT VISITORS: A FLOWER CLUSTER OF THE GREAT RHODODENDRON.

FIG. 105. A COLONY OF TREE-DWARFS AT 5500 FT. ALTITUDE; VIRGIN NORTHERN RED OAK TREES ALL UNDER 30 FT. HIGH, DUE TO THE REPRESSIVE EFFECTS OF LOW TEMPERATURE. SCENE AT FRYING-PAN GAP.

when on the clear days the eye may wander over the merged myriads of broad-leaved trees which collectively, like a great green quilt, cover the vast downward sweeping slopes. One of the most accessible yet wonderful views of this truly great forest is that from Mt. Collins, which may be reached by an easy three or four mile walk or horseback ride from the Newfound Gap auto road. On the Tennessee side the observer may look down a thousand foot precipice onto the misty forest cover far below which continues down the steep slopes to the valley base lying some five thousand feet beneath. Such a view is comparable to that from an aeroplane and from such a scene one may easily realize what the great forest was like in the pioneer days when this same solid leafy blanket spread far and wide over most of the continental eastern uplands.

Before closing this bare outline of the tree composition of the great upland forest it might be of interest to note that only a few trees in it range throughout the state below five thousand feet. Among the more important of these are the beech, white and black oaks, and dogwood (Figs. 106, 107).

The essentials of the great forest habitat have already been indicated. Wherever upland or drained clay, loam and sandy soils of the finer kinds are found, such will carry the plants of medium water requirements. With such soils there is a tendency, as the humus and leaf mold accumulate, for them to become moister, with the result that, in long periods involving centuries, lands that have been in black-jack or pine woodland will go over into oak forest of more moisture loving species, such as the white and black kinds. In damper areas at the bases of the hills, the beech-maple forest may occasionally succeed the oaks. Thus on the medium-watered sites, the same process of plant succession goes on, which we found so strikingly illustrated in the story of the old fields. The vegetational changes described for the weed community continue right on into the forest community until the stable climax is reached in either the oak-chestnut, oak-hickory, or beech-maple type of tree society. Because of the ability of these trees to reproduce in their own shade or partial shade, when a tree dies and falls, the juvenile trees are ready to take its place; the composition of the forest is kept the same. Thus it was that these three major forest types were the principal ones in the piedmont and mountain regions for many thous-

FIG. 106. A COMPETITOR AT THE DOGWOOD SHOW; A TREE FOUND IN CRAVEN COUNTY.

FIG. 107. THE SHOWY PARTS OF DOGWOOD FLOWERS ARE NOT FLOWERS AT ALL; THEY ARE BUT 4 CHANGED LEAVES OR BRACTS, MADE WHITE AND ATTRACTIVE TO INSECTS.

FIG. 108. COTTONWOODS AND WILLOWS OF THE CAPE FEAR AT THE OLD FERRY SITE NEAR ELIZABETHTOWN.

ands of years before the white man came to destroy them. In the coastal plain, because of the high fire incidence, the pines predominated, forming, as has been previously pointed out, a subclimax forest, which means that the succession falls one step short in this region, not going on over into the oak-hickory vegetation, which it surely would do, if fire were prevented for a few thousand years.

In summary, then, we have four major forest communities, three climax and one subclimax, which in the past and in the present (though largely disturbed today) constitute the natural stable or permanent type of vegetation on our landscape.

With this brief introduction to the high woody plants of the great forest which very literally consti-

FIG. 109. THE FLOWERS OF THE VIRGIN'S-BOWER ARE BEST SEEN AGAINST DARK SHADOWS.

tute the "roof trees" of our upland wild flowers, we may proceed to a discussion of a few of the more interesting and rare shrubs and herbs of this community.

SHRUBS: THE WOODY UNDERSTORY OF THE GREAT FOREST

In the plant succession series, shrubs become prominent only in the later stages. That is to say, the greatest development of shrubs is to be found in intimate association with the deciduous climax forest. This holds true not only in quantity but in number of species represented. The richest shrub region, judged from the latter standpoint, is found in the montane portion of the state. In the virgin deciduous forests of the newly opened Smoky Mountain National Park, everyone who visits them, will be impressed by the lux-

FIG. 110. NO SHRUB BLOOMS MORE PROFUSELY THAN THE LAUREL. NOTE ITS FIVE-SIDED FLOWERS AND PINK POLK-A-DOTS. PHOTO BY H. L. BLOMQUIST.

uriant growth of the "little trees." If observing them from a car is not sufficiently appealing, a tramp through them during a rain or when the dew is on them, will make one realize their omnipresence. Under such conditions the only possible comparison is that of a shower bath.

One of the most beautiful shrubs in the world and perhaps the uncontested perennial blue ribbon winner of North Carolina, is the flame-azalea *(Azalea lutea)* of the mountain woods. When Linnaeus gave it the name "*lutea*," meaning yellow, he told only a fraction of the color story for the showy corollas of this remarkable shrub vary widely throughout the range of the warm colors, only stopping where blue comes in to make the reds over into purple. All the possible color tones lying between bright yellows through orange to deepest reds are in the repertory of this plant. The variation appears to be due to both hereditary factors and to the age of the flowers. No more glorious scene in nature may be had than that of dark mountain woods illuminated by the torch-like flowers of the flame azalea. It is little wonder that Michaux named the species "*calendulacea.*"

Among the "runners-up" to the preceding may be mentioned such attractive shrubs as the common laurel (*Kalmia latifolia*, fig. 110) with its polk-a-dot flowers, the flowering raspberry *(Rubus odoratus)*, the shrubby St. John's-worts *(Hypericum)*, of which genus the rare and attractive species "*glomeratum*" may be seen on both Grandfather and Table Rock mountains, two other azaleas *(A. arborescens* and *A. canescens)*, the wild hydrangeas, of which three species may be found, the infrequent two species of syringa *(Philadelphus)*, the silver-bell *(Mohrodendron Carolinum)*, the great rhododendron with its tapering leaf-blade base *(R. max-*

FIG. 111. THE RED-BUD MAKES CERTAIN THAT ITS FLOWERS WILL BE SEEN BY THE INSECTS; THE BLOSSOMS APPEAR WELL BEFORE THE LEAVES. PHOTO BY H. L. BLOMQUIST.

imum), two or three species of wild roses *(Rosa)*, and the early blooming red-bud (*Cercis canadensis*, fig. 111). Also to be mentioned here is the ubiquitous flowering dogwood (*Cynoxylon floridum*, fig. 106).

Shrubs with less showy flowers, yet constituting an important part of the shrub population are such interesting bushes as the wahoo *(Euonymus atropurpureus)*, with its curious fruits which split open and expose the red coated seeds within; the climbing bittersweet *(Celastrus scandens)*, characterized by a drooping cluster of greenish flowers; the related but uncommon Canby's mountain lover *(Pachystima Canbyi)*, a low spreading shrub with narrow leaves and greenish flowers in the leaf axils which produce fruits bearing white-coated seeds; the spice-bush *(Benzoin)* with its highly flavored bark; the common minnie bush

(Menziesia pilosa), which has leaves like an azalea but a flower like a blueberry; the dog-hobble *(Leucothoe Catesbaei),* bearing its compact clusters of flowers underneath the spreading leaves; the mountain pepperbush *(Clethra acuminata),* having slender racemes of small flowers borne from the branch ends; the sweet or strawberry shrubs *(Butneria spp.),* with their very dull purplish highly scented flowers, which have been widely introduced into cultivation; two kinds of hazelnuts which may be easily recognized as the only edible nut bearing shrub; the gooseberries and wild currants *(Grossularia* and *Ribes)* of wide distribution. The witch hazel *(Hamamelis Virginiana)* is remarkable for its late flowers, the blossoms not appearing until September and October.

One small mountain shrub which assumes the aspect of a tree, having a single trunk, is deserving of special mention. This is the buffalo or oil nut *(Pyrularia pubera),* a shrub of restricted distribution which has a simple, ovate leaf difficult to distinguish from those of many other woody plants. Its pendent pear-shaped fruits are very distinctive in that the inedible nut within is not covered at the top by the outer envelope. This shrub also has been reported as being a partial parasite, having its roots attached to those of hemlock trees. In the case of the stand of this shrub best known to the author in the vicinity of Linville Falls, this could not be the case, because of the complete absence of hemlocks from the vicinity.

The shrubs mentioned thus far are but a small per cent of the number which should be treated to present a truly adequate picture of the richness of the mountain, low, woody plant flora. There still remain the blue berries, the sumacs, the species of *Diervilla,* the ninebarks and spireas, the deerberries, the viburnums, and many others. To deal satisfactorily with the shrubs of North Carolina would necessitate another book.

Some Interesting Great Forest Herbs

Asa Gray, an early American botanist visited Paris in 1839 to study the collections there of American plants. In going over the material collected by Michaux in 1788 he came upon a little herb with a leaf like galax but it had only one fruiting body on it indicating a single flower. Gray named it *Shortia galacifolia* but was unable to describe the flower. No one but Michaux had ever found such a plant and when Gray returned with his drawings and description, the search for this rare species was on. Not until 1877 was it found when it proved to be a very attractive wild flower. It is now sold by florists and it may truthfully be said that there are more plants in cultivation than there are wild. D. C. Peattie, a botanist of Tryon, N. C., has introduced it in a wild area in his region.

Another rare plant associated with Gray's name is a mountain species of lily *(Lilium)* named in honor of the American botanist *(L. Grayi).* It is known only from a few stations, the best known of which is that of the mountain alder bald on Roan Mountain. The lily here is found in the somewhat open stand of alder bushes. It is distinguished by the fact that the petals and sepals do not curve backward but funnel-like point forward and slightly outward.

Not so rare, perhaps, as the two species just mentioned but nevertheless of interest because of the restricted distribution in the southern Appalachian mountains, is a remarkably large group of wild flowers which enrich our southern highland flora. The plants of this group are called endemics, since they are only found within a narrow range, the technical manuals referring to their distribution as "Virginia to Georgia in the mountains." Many range into West Virginia and Kentucky.

Only a few of the seventy-eight herbaceous endemics, exclusive of grasses and sedges need be mentioned here. Two trilliums *(T. Vaseyi* and *T. stylosum),* the former with a purplish-brown flower and the latter rose colored, are to be included here. The white "pink" *(Silene ovata)* and the large leaved twin-leaf *(Diphylleia cymosa)* are very characteristic of the mountain slopes. The rose root *(Rhodiola)* and the four-fruited stone crop *(Tetrorum)* are of local occurrence in the high mountains. Other interesting endemics are: the ginger-leaved grass-of-Parnassus *(P. asarifolia),* the giant leaved foam-flower *(Tiarella macrophylla),* aconite saxifrage *(Therophon aconitifolia),* which reaches the height of three feet, the high mountain saxifrage *(Hydatica petiolaris),* both species of false goat's beards *(Astilbe),* the mountain avens *(Sieversia radiata),* the four species of bush pea *(Thermopsis),* the very common galax *(Galax aphylla),* the extremely rare *Shortia,* the two phloxes, *P. Brittonii* with white flowers and *P. acuminata,* a purple-flowered one with decurrent leaves, four mountain mints *(Koellia),* the coarse stoneroot *(Collinsonia tuberosa),* the rose-colored turtle-head *(Chelone Lyoni),* two beard-tongues *(P. Smallii* and *P. canescens),* the creeping bluet *(Houstonia serpyllifolia),* four wild gingers

(Hexastylis Shuttleworthii, heterophylla, Ruthii and *rubricinctum)*, the delicate blue bell *(Campanula flexuosa)*, the angled stem golden-rod *(Solidago spithamea)*, two or three species each of aster, rosinweed, coneflower and sunflower, the beautiful *Coreopsis latifolia* and three species of ragwort *(Senecio)*. Especially interesting is the occasional occurrence of the lily-of-the-valley *(Convallaria)*, the same species which has found its way into thousands of gardens.

Most of these endemics are strictly confined to the mountain area and are there frequently rare and of local occurrence. A few like galax occur sporadically far to the east. One such station lies three miles south of Cary in Wake County where the galax forms a solid sod on a steep northward facing slope under a grove of hemlocks and chestnut oaks, mountain trees which also have held on here from an earlier period.

In contrast to the endemics just mentioned, many wild flowers of the great forest community are familiar to every one. Who, as a child, on the springtime excursions to the woods, has not been thrilled with the experience of picking large handfuls of spring beauties, bluets, dog-tooth violets, hepaticas, and buttercups? Who does not know from early years the mayapple, trailing arbutus (accent on first syllable), cranesbill, wild ginger, saxifrage, and the trilliums? And how much one has missed who is not on a speaking acquaintance with the foam flower, yellow star-grass, the two dwarf irises, the partridge berry, bellworts, stone-crops and pennywort. To know exactly in the woods where Solomon keeps his seal and where the Indian finds his pipe is very worth while knowl-

FIG. 112. SIMULATING A TROPICAL LUXURIANCE, THE HAY-SCENTED FERN MASSES ARE FREQUENT IN MOUNTAIN WOODS. PHOTO BY H. L. BLOMQUIST.

edge—if one is still youthful enough to appreciate such valueless information.

Most of the above familiar plant friends are spring bloomers. Another list could be given for the summer, though it is to be suspected that these would not be so well known. Late summer brings the low ebb in wild flowers, which gives way to a final renewed wave of color with the coming of the fall-time goldenrods and asters, genera with which everyone is acquainted.

Just why nature chose her little folk, the herbs, to be the chief showy flower bearers, the shrubs next, and the trees last, may not be certainly known, but it is a fact nevertheless, that without these the humblest residents of the forest, the woods would lose their principal charm, and gardeners would be deprived of one of their richest sources of floral loveliness.

Soil areas as moist as are the mountain slopes would naturally be expected to carry a complete line of ferns—those beautiful delicate-leaved plants which are always associated with cool shady places. And such is the case. Ferns live there in riotous profusion. Large open areas locally may be dominated by the sweet-smelling hay-scented fern (*Dennstaedtia punctilobula,* fig. 112). In shadier places abundant displays of the Christmas fern *(Polystichum acrostichoides),* the silvery spleenwort *(Athyrium thelypteroides),* the marginal shield fern *(Dryopteris marginalis),* the maiden-hair fern *(Adiantum pedatum),* and many others are to be noted. The curious walking fern *(Camptosorus rhizophyllus)* is occasionally encountered, the plants spreading by the rooting leaf tips. Numerous small ferns of rock crevices add their mite of leafy beauty to the rock setting. A few of such are the mountain woodsia *(W. obtusa),* the lip-ferns *(Cheilanthes),* the polypody fern *(Polypodium vulgare),* mountain and maiden-hair spleenworts *(Asplenium montanum* and *A. Trichomanes).* Much as a florist uses fern leaves in floral backgrounds, so our great forest ferns furnish an exquisite setting for the wild flowers in and near them. Their leaf forms are generally more attractive than those of the higher plants, so that it may truly be said that a vegetation increases in beauty as its fern numbers increase. Judged by such a criterion our mountain region will rank high.

Going below the ferns in the plant scale, into the mosses, mushrooms, and others we may simply say that the great forest areas of North Carolina are extraordinarily rich in species and numbers of individuals in all of these lower plant groups. How much luxuriant moss beds can add to the woodland scene! And how amazingly rich are the mountain slopes in these miniature thickets of the vegetational world! And as to mushrooms, the writer once heard the late Horace Kephart tell of an amazing number of these plant umbrellas he once observed on a half acre in the Smoky Mountains. All students of the fleshy fungi recognize the western moist mountain areas as one of the very richest collecting grounds in North America. If one knows his mushrooms as well as his onions, he need not starve in the mountain region.

In summary of this chapter let us again make it clear that in the area covered by the great forest community, we have the medium soil water conditions and moderate temperature which favor the highest development of a vegetation. The vegetation always responds to the environmental conditions. The maximum appearance of these favorable conditions in the United States is reached in the well drained areas of the southeastern United States, where the rainfall quantity and distribution are most favorable for plant growth. This has resulted in the appearance in this region in past ages of the great deciduous forest and its associated shrubs and herbs, a forest which, were it not for fire, would have also covered the uplands of the coastal plain.

CHAPTER X

CHRISTMAS TREE LAND: THE BOREAL FOREST OF OUR HIGH MOUNTAINS

OUR CANADIAN friends may not recognize it but all of Canada is by no means beyond the northern boundary of the United States. Of course we are speaking climatically and we may do so because of the rule that high altitude and latitude give us similar climates. Whether we go up or go north, it just gets cold in either case. Thus along the entire range of our eastern mountains, here and there the summits become high enough to be sufficiently cold to suppress the oaks and other broad leaved trees. This check upon the great forest makes possible the advent of the same kind of Canadian forest which is found north of the deciduous tree region.

And this forest, romantically enough, is made up of genuine Christmas trees (Fig. 113). In Europe for centuries the balsam firs and spruces have been used at Yuletide; no other trees are somehow equal to these for displaying the candles or, in the modern period, the electric colored lights.

The vast stretches of fir forest in Canada grow at low altitudes; on the Newfoundland coast, the trees come down to the sea. But in North Carolina the same climatic conditions are not met until an altitude of fifty-five hundred feet or greater is reached. Since the state has forty peaks over six thousand feet, it is

FIG. 113. CHRISTMAS TREE LAND: A GLIMPSE OF CANADA IN NORTH CAROLINA: NOTE GRASS BALD AND SHRUB BALD IN DISTANCE, RIGHT AND LEFT RESPECTIVELY. FOREGROUND TREES CHIEFLY FRASER'S BALSAM. VIEW ON ROAN MOUNTAIN.

readily seen why by far the largest areas of boreal forest in the Southern Appalachians are included within the boundaries of North Carolina.

It is astonishing for the inexperienced, in ascending the mountains to change forests within half a mile, or on steep slopes the distance may be even less (Fig. 114). From certain lookout points the observer can look over the tree tops at the high altitudes and see the transition region where the deciduous trees become dwarfed and disappear with the accompanying appearance and increase of the balsams and spruces. It is difficult to make oneself believe that in that short distance which leads from one pure forest type to an entirely different one, a complete change of climate

FIG. 114. WHERE THE CLIMATE CHANGES IN A HALF MILE WALK; THE TRANSITION FROM HARDWOODS TO SPRUCE AND BALSAM.

occurs. This change, insofar as the vegetation is concerned, is greater than the difference between the climate of the mountain slopes and that of the coastal plain.

The most significant change among the number of climatic factors present is the change in temperature. So sudden and so severe is the onset of freezing temperatures in both the fall and spring that the broad-leaved trees are not able to stand such extreme changes, with the result that winter killing and growth repression are so great that these trees become insignificant as competing units of the vegetation.

The boreal trees, it has been discovered, are adapted to the extreme cold because of the presence in their cells of certain fatty substances which are made

FIG. 115. LIKE A PORCUPINE THE ISOLATED WIND BLOWN RHODODENDRON STANDS READY FOR ANY ATTACK.

from starch manufactured by the leaves. These fatty or oily materials not only depress the freezing point but they have a high water holding capacity which prevents the loss of water under the pull induced by crystallizing water outside of the cell. Modern theories of freezing injury hold that such injury is induced largely by excessive loss of water from the cells and is to be compared to what happens to a plant when it wilts.

However, without attempting here to go far into technicalities, we may be very certain that the hardiest native plants of our state are naturally those which grow at the highest altitudes where temperatures of twenty-five and thirty degrees below zero are not uncommon and where the growing season is confined to a little over three months of the year. Not only the trees but numerous shrubs and many herbs are to be included in this class of hardiest plants. On certain high summits, which are exposed to all the high desiccating winds, the fir trees give evidence of repression and show a one-sided growth. The rhododendron becomes very compact (Fig. 115). If our mountains were high enough, as may be observed in the western ranges, an altitude would be reached at which the climate would be too severe for even the balsams and the spruces. They fall short of that elevation by a few thousand feet.

Typical High Mountain Plants

Since the number of common plants of the boreal forest is very restricted, we shall make mention of a few of the more interesting ones.

The balsam *(Abies Fraseri)* may be distinguished from the spruce by its flattened leaves which, when they drop from the branch, leave a circular slightly depressed scar. The spruce leaves are four-angled and, upon shedding, leave a bit of the base attached to the twig. Thus the older spruce twigs are made rough by these innumerable little processes sticking out all over them. Our southern Appalachian balsam is slightly different from the Canadian balsam *(Abies balsamea)*.

Of the spruces we have two. The black spruce *(Picea Mariana)* with its slightly heavier leaves, has the aspect of a balsam so that the mountaineers call it the "he-balsam," the true balsam being the "she-balsam." Just how the sex was decided in this case would be a matter of much interest, for both trees are, of course, bisexual, bearing pistillate and staminate

FIG. 116. AMID THE SILENT SPRUCES ON THE OLD TRAIL TO INDIAN GAP.

cones on the same tree. The red spruce is not as attractive a tree as the black. The spruces are more frequent at the lower altitudes within the boreal forest, so that when entering the mountain peak forests from any side, the spruces will at first be very abundant (Fig. 116) but give way to the dominance of the balsam at the higher altitudes where in many peaks the stand is a practically pure one.

These three Christmas trees constitute the body of the cold loving forest. All other woody plants play a very minor rôle in this community. The mountain ash *(Sorbus Americana)*, a small tree with its pinnate leaves and its flat-topped cluster of bright red little apples, is certain to be encountered along with the shrub known as the hobble-bush *(Viburnum alnifolium)*. This plant also has a flat-topped inflorescence of white flowers in which the marginal flowers display

FIG. 117. EARLY JUNE, RHODODENDRON FLOWERS AND THE HIGH MOUNTAINS MAKE AN UNFORGETTABLE TRINITY IN EXPERIENCE.

much larger corollas. The high mountain rhododendron *(R. Catawbiense)* is scattered widely in the more open forests. This shrub, however, expresses itself best in the open (Fig. 117), and hence a fuller discussion of it will be given later in the treatment of the balds. The wild red cherry *(Prunus Pennsylvanica)* is scattered in the open places contributing seeds to the soil which, in case of destruction of the evergreen forest by fire, will germinate and contribute to the successional community. The mountain maple *(Acer spicatum)* ranges into the high mountain community.

Of especial interest in the balsam-spruce woods is the forest floor. Best developed where the land is not steep, is the beautiful tree-moss *(Hylocomium splendens)*, which grows so thickly over the ground as to make a thick turf. From the surface arise the stems and branches, simulating in miniature the forest above it. Like sphagnum moss it holds water so that old accumulations of it on level mountain tops or ridges

FIG. 118. AN EXPLORATION OF THE GRASSY BALD ON ROAN. NOTE SHARPNESS OF TRANSITION TO SHRUBS AND TREES.

are commonly as soggy as the soil of the eastern bogs. The impression always left with the visitor to the "black forest" mountain tops is one of dampness. The deep shade of the closely spaced trees helps the hylocomium moss exceedingly in its work of holding the water at the mountain top. So wet are some of the high, yet level areas, that peat has accumulated, which means that a virtual moss-choked pond caps these summits. One may properly speak of our mountain-top bogs.

In midsummer the deep shade is illuminated by thousands of flowers of the purple wood sorrel *(Oxalis Acetosella)*, which push their bright faces above the moss. Surely no wild flower ever had a finer setting in which to display its loveliness than the feathery tree-moss furnishes for the oxalis. A few other flowers also enjoy this same privilege. The delicate little white violet *(Viola blanda)* and the broad bladed *Clintonia borealis,* with its greenish-yellow flowers and its blue berries, are certainly to be mentioned. The tall green hellebore with its large, oval, sessile leaves occasionally attracts attention.

The Mountain Balds

No plant communities in the state have been more puzzling to the ecologist than those that come under the designation of mountain balds (Fig. 118). These are the high mountain areas which are treeless, the surface now being covered by either an herbaceous or a shrub vegetation or these more or less mixed. Since the forest is found much higher, being present on the two highest mountains, Mitchell and Clingman's Dome, we may be sure that the balds are not to be compared with the low mountain tundra of western mountains, a grass and shrub vegetation found only above the timber line. The eastern mountain balds have a very different story behind them. But as we have already indicated that story is far from being fully known.

All sorts of theories have been imagined to account for these forest-free areas. One student believed that they were all so located that ice storms were more frequent in their respective sites with consequent continual breaking of tree limbs in winter thus making it impossible for the trees to compete with the smaller vegetation. Soil acidity, wind, fire, parasites, almost every important ecological factor has been called upon to contribute its share in explaining how a bald came into existence and how it is maintained.

FIG. 119. A REMARKABLE SEDGE (CAREX) BALD ON MT. STERLING. PHOTO BY H. L. BLOMQUIST.

When the earliest settlers came into the mountains, the balds, certainly the largest ones, were already there. Their age is to be measured in centuries. Why this hesitancy to go back to forest when forests are all around them? In all of the rest of the state when a treeless area is left undisturbed it is but a matter of from five to twenty years until the pines begin to take it, and on most sites, if no fire comes, the oak-hickory or oak-chestnut forest will follow the pines. It may thus be seen that the balds are all out of joint with the rest of our vegetation; they ought to disappear but they don't.

One peculiar characteristic of them is that they vary widely in composition. The herbaceous or grassy balds (Fig. 118) show different grasses dominant in different ones. In others sedges play a prominent rôle, such as one reported by Dr. H. L. Blomquist on Mt. Sterling (Fig. 119). Locally in the herbaceous type, the hair cap moss will rule remarkably large

FIG. 120. NO GARDENER COULD SURPASS THE WORK OF WILD NATURE IN THE HIGH MOUNTAINS IN CREATING A PARK-LIKE ASPECT.

FIG. 121. THE MOUNTAIN ALDER *(Alnus viridis)*, ONE OF THE RAREST OF OUR SHRUBS. IT IS BELIEVED TO OCCUR ONLY AT THE ONE STATION ON ROAN MOUNTAIN.

FIG. 122. A SHRUB BROBDINGNAGIAN AND LILLIPUTIAN MEET ON A MOUNTAIN BALD: RHODODENDRON AND SAND MYRTLE.

areas. In rocky areas of rounded tops, the mountain cinquefoil *(Potentilla tridentata)* and the exquisite little bluet *(Houstonia serpyllifolia)* are to be expected.

The shrub balds, though most frequently represented by rhododendron (*R. Catawbiense,* fig. 120), nevertheless show a wide variation in subdominance. On Roan Mountain is a shrub bald of the green alder (*Alnus Alnobetula,* fig. 121). On exposed cliffs the low box-like shrub, sand myrtle (*Dendrium prostratum,* fig. 122) is frequent. Certain balds exhibit a dominance of the beaked hazelnut *(Corylus rostrata)* with azalea *(A. arborescens)* and shrub honeysuckle *(Diervilla sessilifolia)* as sub-dominants. Locally the high bush blueberry *(Vaccinium corymbosum)* may be abundant. At the lower altitudes, laurel *(Kalmia latifolia)* in a suppressed condition makes a glowing patch of color when in bloom.

Coming back to the questions of the origin and maintenance of these treeless areas, no positive statements will be made in this book; the writer has made no intensive studies and not until these are made and continued over many years, will ecological science begin to solve the riddle of the balds. In the meantime working hypotheses may be held. A few of these in the mind of the writer are: that fire, not one but a number in relatively close sequence, originally eliminated the forest, making possible the entry of the bald species; that the high altitude conditions, strongly favor the shrubs and grasses in their competition with the tree species; that the thick grass may actually encroach on rhododendron without the interference of fire; that the forest may regain the thickly covered bald areas only by moving in on them laterally, a process so slow as to be difficult to demonstrate.

Whatever the explanation of the mountain balds may be, plant lovers should be thankful for the wonderful development on many of them of the extremely beautiful high mountain rhododendron *(R. Catawbiense)* already mentioned. If, at nature's shrub show, we give the flame azalea the blue ribbon, the rhododendron should be awarded the red one. Whoever visits a rhododendron bald in early June will find himself in the presence of one of the finest natural scenes to be enjoyed anywhere in the world. Where a bluet-dotted grass area is present to give the shrubs a park-like setting, the rhododendron with its luxuriant masses of purple flowers makes an unforgettable picture. One such scene was enjoyed by the writer and a party on the

FIG. 123. WHERE RHODODENDRON TURNED INTO DOLLARS AND ROCKS. IT WILL TAKE TWO THOUSAND YEARS OR MORE TO REPLACE NATURALLY THIS SHRUB WHERE THE SOIL IS REMOVED WITH THE PLANTS.

Craggy mountains, at which time huge white cumulus clouds were present to add to an already exquisite landscape. Like a broker giving a tip on the market, the author would here drop a tip to the reader who is sensitive to floral beauty, that he some year in early June give himself the treat of seeing the high mountain rhododendron celebrate the coming of spring. As in the case of a successful financial coup, he will never regret the experience.

Before leaving this brief mention of a notable shrub, a word of warning should be given relative to the rhododendron balds. So great has been the demand for this most beautiful plant, that extensive areas have already been denuded by the nurserymen who, working with a thin soil, leave naught behind but the bare rock surface (Fig. 123). It will take thousands of years for nature to rebuild the soil in such places. It must someday become the business of the state to prevent excessive destruction of such a vegetation, for like many other natural resources, the supply is limited and cannot be readily replaced.

This limited discussion will be sufficient to make clear the interesting problem of the high, open mountain localities. In any event all mountain lovers are glad that the balds are there. By the introduction of such delightful diversity into the mountain scenery, they have enhanced extremely the magnificent beauty of the lands of the sky.

CHAPTER XI

NATIVE WILD FLOWERS FOR THE GARDEN

Our title for this chapter is a misleading one, for all garden plants have been derived from wild ancestors. Our finest cultivated flowers are, like man himself, just a few steps removed from the primitive state. In the wild places of South Africa today one may collect the geranium, freesia, and gladioli of the loveliest hues. The Russian children collect tulips as ours do buttercups and hepaticas. The jonquil, daffodil, snowdrop, and forget-me-nots are all native in Europe, while the sweet pea originated only in Italy. Our common house begonias grow luxuriantly wild in Brazil along with fuchias, while heliotrope is confined as a wild flower to Peru and petunias to Argentina. Many of our primulas beautify the slopes of the Himalayas. Cosmos is one of the most beautiful wild flowers of Mexico. So it is with all the rest—somewhere members of the various species are carrying on without the aid of human gardeners.

We may be very sure that when we add some of our native wild flowers to our gardens we are in no sense whatsoever depreciating the garden's value. On the other hand, we are but enhancing the joy of the gardening process by adding the many easily obtained examples of plant beauty from our own wild places, and in so doing we are giving adequate representation to our own flora in that cosmopolitan area we call the garden.

From the information given above concerning origins, it may readily be seen that there is a fascinating geographical side to gardening. With the aid of the manuals to the cultivated plants, any gardener may learn the native country of his various species. And how much pleasure it will add to the development of the garden plot, to realize that all the continents of the world have contributed to the beauty of the garden space! Through the long, long ages of recent geological time the plants in all parts of the world have been progressing, and because of geographical isolation, the various land areas have come to have very different floras. But in our gardens we break down all such isolation and turn ourselves through our plants, into true cosmopolites.

In the suggestions to follow concerning the use of North Carolina plants, let it be made clear at once that ours is a restricted list. Every gardener should feel free to bring into his garden any of our native things which strikes his fancy. Whether the name is known or not, bring it in.

One of the most fallacious of prevalent ideas is that we do not know an animal or a plant until we have learned its name. If one is a good observer and sensitive to beauty, he may exhaust the possibilities of any new and strange plant and go his way; the name, if he finds it out later, is merely a convenient handle by which he may present his knowledge and appreciation to others. The name may come last or even not at all; the knowledge of and aesthetic joy in the plant itself is the all important matter.

In keeping with the whole tenor of this book the plant lists are presented in relation to the type of garden in which the plants are to be used. These kinds of garden areas, differing of course in the all important character of the average soil water content, are in reality but imitations of the natural soil habitats to which the various plant communities are adapted. This ecological principle is at the very foundation of successful gardening, and by the expert may be carried out to the extent of slight modification of local habitat to make the conditions more favorable for a particular species. For plants in the same community differ among themselves and choose what one writer has called "ecological niches." Such a "niche" to define it, would mean that particular cluster of conditions around a plant which make for it an optimum habitat. Thus the gardener's problem is a very clear one; viz., to create these "ecological niches" which are occupied by his various plants. But how to do it, in all cases, is a task that is still baffling the best brains in the horticultural art. For the wild plants, we may gain a big hint, of course, from the native habitat. Note where the plants to be used are making their best growth and then not only use those plants but study the soil habitat closely and develop as nearly as possible, similar conditions in the garden. For a very restricted development of a special plant, it may prove profitable to transplant it by taking it up in a large amount of its native soil. In its new site, the fundamental water relations may be changed in too great a degree interfering with success, but ordinary judgment in a matter of this sort should avoid such an error. The motto

"Follow Nature" is a good one in solving the problem of introducing native plants into the garden areas.

This is not a book on landscape architecture, but it is desirable to make another point in that field before proceeding to the suggested flower lists. Our native plants lend themselves much more to the naturalistic type of garden than the conventional or formal one. This fact is fortunately in harmony with the recent trend in landscape practice, which is definitely toward the nature-imitation ideal.

It makes one happy to realize that this goal is favored by all of the factors: greater beauty is achieved, a large and diverse amount of material is available in our wild things, and last but not least the expense is at a minimum. A "natural" garden thus may be created by almost anyone. With such a profusion of material to draw from, North Carolinians could, if they would, easily turn their inhabited areas into localities which would be known far and wide for their floral attractiveness. The day might even come when it would be known as the "Garden State."

Lists of our native herbs which should prove desirable for the various types of gardens may be of value not only in their direct suggestions but also may furnish an ecological reference sheet which can guide the gardener as to the proper location of material collected. A few plants will be found in more than one list since they have an unusually extended range of habitat. The common names will be followed only by the generic scientific names in most instances.

FOR THE SUNNY ROCK GARDEN

Alum root *(Heuchera)*
Beargrass *(Yucca)*
Bird's foot violet *(Viola pedata)*
Blazing stars *(Laciniaria)*

NATIVE WILD FLOWERS FOR THE GARDEN

Bluets *(Houstonia)*
Blue curls *(Trichostema)*
Bunch flower blazing star *(Chamaelirium)*
Butterfly peas *(Clitoria, Bradburya)*
Butterfly weed *(Asclepias)*
Cactus *(Opuntia)*
Columbine *(Aguilegia)*
Cow-wheat *(Melampyrum)*
Creeping twin-fruit *(Dichondra)*
Corn salad *(Valerianella)*
Dwarf false dandelion *(Adopogon)*
Dwarf iris *(Iris critata* and *I. verna)*
Elephant's foot *(Elephantopus)*
Everlasting *(Antennaria)*
Fame flower *(Talinum)*
Flowering spurge *(Chamaesyce)*
Fog-fruit *(Phyla)*
Frost-weed *(Helianthemum)*
Gerardia *(Gerardia)*
Gaillardia *(Gaillardia)*
Golden aster *(Chrysopsis)*
Gromwell *(Lithospermum)*
Golden saxifrage *(Chrysosplenium)*
Indian physic *(Porteranthus)*
Jewel weed *(Impatiens)*
Loud speakers *(Marshallia)*
Lousewort *(Pedicularis)*
Moss pink *(Phlox subulata* and *Hentzii)*
Mountain avens *(Sieversia)*
Mountain bluebell *(Campanula divaricata)*
Painted cup *(Castilleja)*
Pencil flower *(Stylosanthes)*
Pennyroyal *(Hedeoma)*
Pineweed *(Sarothra)*
Sandhill gentian *(Dasystephana Porphyrio)*
Sandhill wild chickweed *(Alsinopsis Caroliniana)*
Saxifrage *(Micranthes)*
Scaly seed *(Spermolepis)*
Scarlet flowered bean *(Erythrina)*
Shortia *(Shortia)*
Small-leaved pyxie *(Pyxidanthera)*
Southern rock orpine *(Diamorpha)*
Spring beauty *(Claytonia)*
Spring green and gold *(Chrysogonum)*
Stiff-leaved aster *(Ionactis)*
Stonecrop *(Sedum)*
Trailing arbutus *(Epigaea)*
Twin-fruit *(Amsonia)*
Turkey-beard *(Xerophyllum)*
White-topped aster *(Sericocarpus)*
Wild pinks *(Silene)*
Wood sorrel *(Xanthoxalis)*

FOR THE SHADED GARDEN INCLUDING WELL SHADED ROCK GARDENS

Alexanders *(Zizia)*
Alum root *(Heuchera)*
Baneberry *(Actaea)*
Bellwort *(Uvularia)*
Bishop's Cap *(Mitella)*
Black cohosh *(Cimicifuga)*
Bloodroot *(Sanguinaria)*
Blue cohosh *(Caulophyllum)*
Blue darts *(Psoralea)*
Bunch-flower blazing star *(Chamaelirium)*
Bush pea *(Thermopsis)*
Cardinal flower *(Lobelia)*
Carolina wind-flower *(Anemone)*
Clinton's lily *(Clintonia)*
Closed gentian *(Dasystephana Andrewsii)*

Columbine *(Aquilegia)*
Cowslip *(Dodocatheon)*
Crane-fly orchis *(Tipularia)*
Dog tooth violets *(Erythronium)*
Dwarf iris *(Iris cristata* and *I. verna)*
False goat's beard *(Astilbe)*
False wintergreen *(Pyrola)*
Foam flower *(Tiarella)*
Fringed gentian *(Anthopogon)*
Galax *(Galax)*
Giant chickweed *(Alsine pubera)*
Goat's beard *(Aruncus)*
Greek valerian *(Polemonium)*
Hellebore *(Veratrum)*
Hepatica *(Hepatica)*
Jack-in-the-pulpit *(Arisaema)*
Ladies' slipper *(Cypripedium)*
Larkspur *(Delphinium)*
Leopard's-bane *(Arnica)*
Lily-of-the-valley *(Convallaria)*
Lupine *(Lupinus perennis)*
May apple *(Podophyllum)*
Monkshood *(Aconitum)*
Partridge berry *(Mitchella)*
Pennywort *(Obolaria)*
Purple wood sorrel *(Oxalis)*
Rattlesnake plantain *(Peramium)*
Sabbatia *(Sabbatia)*
Shortia *(Shortia)*
Showy orchis *(Galeorchis)*
Solomon's seal *(Salomonia)*
Trilliums *(Trillium)*
Turtle-head *(Chelone)*
Twin-fruit *(Amsonia)*
Violets *(Viola)*
Virginia blue bell *(Mertensia Virginica)*
Water pennywort *(Hydrocotyle)*
Wild geranium *(Geranium)*
Wild ginger *(Asarum, Hexastylis)*
Wild phloxes *(Phlox)*
Wild pinks *(Silene)*
Wild spikenard *(Vagnera)*
Wild yam *(Dioscorea)*

FOR THE MARSH AREA

American lotus *(Nelumbo)*
Arrow-heads *(Sagittaria)*
Atamasco lily *(Atamosco)*
Bartonia *(Bartonia)*
Beautiful marsh mint *(Macbridea)*
Button snakeroot *(Eryngium)*
Cardinal flower *(Lobelia)*
Dragon-head *(Physostegia)*
False loosestrife *(Ludwigia)*
False pimpernel *(Ilysanthes)*
Germander *(Teucrium)*
Golden club *(Orontium)*
Golden saxifrage *(Chrysosplenium)*
Green arrow-arum *(Peltandra)*
Hedge hyssop *(Gratiola)*
Iris *(Iris prismatica, fulva, versicolor, Caroliniana)*
Lizard tail *(Saururus)*
Long-stalked false loosestrife *(Ludwigiantha)*
Loosestrife *(Lysimachia* and *Steironema)*
Marsh blue bell *(Viorna crispa)*
Marsh buttercups *(Ranunculus delphinifolius* and others)
Monkey-flower *(Mimulus)*
Marsh gerardia *(Gerardia fasciculata, paupercula)*
Marsh green orchis *(Perularia scutellata)*
Marsh loosestrife *(Ammania)*
Marsh marigold *(Caltha)*
Marsh purslane *(Isnardia)*

NATIVE WILD FLOWERS FOR THE GARDEN

Marsh skullcap *(Scutellaria lateriflora)*
Marsh St. John's-wort *(Triadenum)*
Marsh tick-trefoil *(Coreopsis)*
Pickerel weed *(Pontederia)*
Primrose willow *(Jussiaea)*
Purple-head waterweed *(Sclerolepis)*
St. John's-wort *(Hypericum adpressum* and others)
Thread plant *(Burmannia)*
Tooth-cup *(Rotala)*
Water cress *(Roripa)*
Water hoarhound *(Lycopus)*
Water parsnip *(Sium)*
Water pennywort *(Hydrocotyle)*
Water pimpernel *(Samolus)*
Water willow *(Dianthera)*
Yellow-eyed grass *(Xyris)*

AQUATIC PLANTS FOR POOLS

Bladderwort *(Utricularia)*
Carolina water-shield *(Cabomba)*
Duck weeds *(Lemna)*
Dwarf duckweed *(Wolffia)*
Feather-foil *(Hottonia)*
Floating-heat *(Limnanthemum)*
Frog-leaf *(Brasenia)*
Greater duckweed *(Spirodela)*
Horned pondweed *(Zannichellia)*
Hornwort *(Ceratophyllum)*
Mermaid weed *(Proserpinaca)*
Parrott's feather *(Myriophyllum)*
Pondweeds *(Potamogeton)*
Spatterdock *(Nymphaea)*
Tape grass *(Vallisneria)*
Water crowfoot *(Batrachium)*
Water fern *(Azolla)*
Water lily *(Castalia)*
Water starwort *(Callitriche)*
Water weed *(Philotria)*

FOR THE SUNNY GARDEN BOG

Variety is surely one of the spices of gardening. Nothing could more successfully introduce a new note in the garden area than a small bog with its beautiful flowers blooming throughout the summer. The year 1931 was a rock garden year. Perhaps some year soon to come will be a bog year, when every garden of any consequence will add a small bog.

To make the proper soil habitat for such plants, it will be necessary to create a high water table for extended periods by placing the soil in a concrete basin. This should be at least a foot in depth and filled preferably with bog soil. If no dark, heavy soil is available a large amount of dried sphagnum moss (florist packing material) may be added to loam soil and after a period of flooding, such a soil will readily grow the acid loving plants. The rains ordinarily should keep the soil wet enough, but in the longer drouths, water should be added from time to time.

The bog garden will not need much weeding since most of the common garden pests find the wet soil a very unhealthy environment for their roots.

The following plants may all be found in abundance on the grass-sedge bogs or savannahs of the lower coastal plain.

Blue hearts *(Buchnera)*
Bog blazing star *(Laciniaria Wellsii)*
Bog Coreopsis *(Coreopsis gladiata, falcata)*
Bog false violet *(Pinguicula pumila)*
Bog fly poison *(Tracyanthus)*
Bog fringed orchid *(Blephariglottis)*
Bog iris *(Iris tripetala)*
Butterworts *(Pinguicula)*
Chain fern *(Woodwardia)*
Clustered bluets *(Oldenlandia)*
Colic root *(Aletris farinosa)*
False asphodel *(Tofieldia)*
Gerardia *(Gerardia purpurea)*
Grass pinks *(Limodorum)*
Ladies tresses *(Ibidium)*
Meadow beauty *(Rhexia)*
Night-nodding bog dandelion *(Thyrsanthema)*
Pyxie *(Pyxidanthera barbulata)*
Rayless flat-topped golden rod *(Chondrophora)*
Rough-stemmed bunch flower *(Triantha)*
Sabbatia *(Sabbatia dodecandra)*
Savannah iris *(Iris tripetala)*
Savannah milkworts *(Polygala lutea, ramosa)*
Savannah white orchid *(Gymnadeniopsis nivea)*
Snake mouth orchid *(Pogonia)*
Spider lily *(Hymenocallis)*
Star flower *(Pleea)*
Sundew *(Drosera)*
Tall bunch flower *(Zygandenus)*
Trumpets *(Sarracenia flava, S. minor)*
Venus' fly-trap *(Dionaea)*
White bracted sedge *(Dichromena)*

PART II

THE HERBACEOUS WILD FLOWERS OF THE NATURAL GARDENS

SECTION I

ARTIFICIAL KEY TO THE GENERA OF HERBACEOUS WILD FLOWERS OF NORTH CAROLINA

In this book on wild flowers, habitat considerations have been given a prominent place. Such an emphasis will be of significance in relation to the keys which follow, for the author, despite many objections of which he is fully aware, believes more will be gained than lost in such a procedure. The primary division of the herbaceous wild flowers has been based upon the fundamental kinds of soil habitat which as ecologists are well aware, carry very distinctive vegetations. The first part of this book presents a description of these habitats and their respective floras which should help the reader very much in his development of a habitat consciousness which will more than double his interest in plant life. It will be very desirable for those who live in the coastal plain, to carefully read the account of the eastern communities for it is in this region of the state that the larger number of fundamental habitats and vegetations occurs.

The greatest single gain in organizing our keys on the habitat basis lies in the fact that the collector, if he expects to use this book for identification purposes, must note the outstanding habitat and correlated vegetation when he collects his specimen. Thus he will be forced to think ecologically and in so doing will find himself in line with the modern movement in biology which attempts to understand the organism in relation to its environment and to organize the masses of organisms into natural communities isolated by fundamental differences in the habitats.

Another gain of particular significance in a state like North Carolina, which is characterized by such a wide range in both soil and climatic conditions, lies in the fact that the reader need only give attention to those habitats and communities which occur in his vicinity. Thus instead of being forced to wade through a very long general key, he may use the more restricted special keys. For the piedmont and mountain regions, the single extensive key to the upland plants, will be used for the most part, since there is so little development in these areas of marsh or aquatic life. For the inhabitants of the coastal plain, and especially the lower coastal plain, the problem is more complex, but let us hope this will but add to the interest in the matter of finding out what their plant neighbors are.

Directions For the Use of Key and Manual

We have classified the wild flowers into seven of the eleven major communities in North Carolina. The old field and boreal forest plants have been included in the great forest key. Of the weeds only the showier-flowered ones have been added. The swamp forest and shrub-bog or pocosin areas present but few herbaceous wild flowers and these because of similarity of habitat, have been included in the key to the savannah species.

Suggestions for the trial of other keys if the one first used fails, will be given with the separate keys.

Before proceeding it will be necessary for the reader to familiarize himself with the few terms illustrated in Figs. 124-127, which are used to distinguish the various principal kinds of flowers, inflorescences, leaf forms and leaf arrangements.

A brief summary of flower structure will be in order here, to aid the beginner in flower study. There are commonly 4 sets of parts to a typical flower: the outer calyx made up of

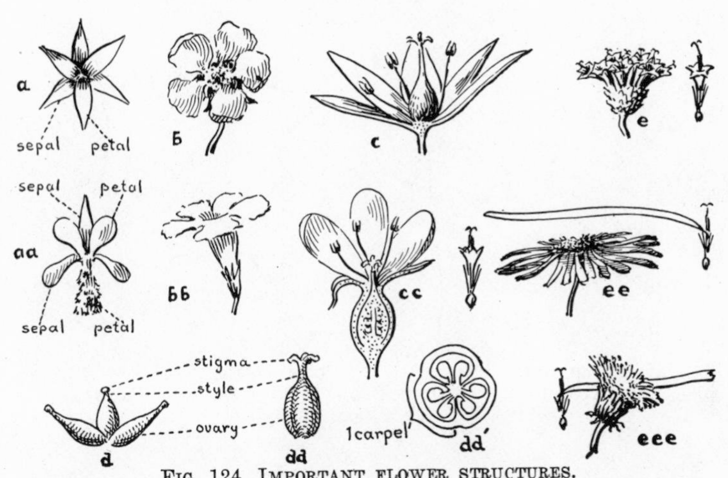

FIG. 124. IMPORTANT FLOWER STRUCTURES.

a, Flowers radiate: Such flowers may be divided into equal halves by 2 or more planes. Examples: lily, buttercup, bluet. *aa,* Flowers bilateral: Flowers may be divided into equal halves by 1 vertical plane only. Examples: pea, violet, orchids.

b, Petals separate. *bb.* Petals united. Petals may be barely united at base as in the milkweed, or completely joined as in the morning glory.

c, Sepals and petals borne below the ovary. Examples: lily, chickweed. *cc,* Sepals and petals borne above the ovary. Examples: iris, orchids, evening primrose.

d, Carpels separate; *dd,* carpels united; *dd',* ovary in cross section showing 3 carpels united.

e, Flowers in a head. Head with corollas all tubular, the flower bases surrounded by the involucre of small overlapping bracts arranged in many series. Single, tubular flowers at right. Examples: Thoroughwort, blazing-star. *ee,* Head with central corollas tubular, marginal flowers strap-shaped. Tubular, type left, strap-shaped, right. Examples: Aster, sunflower. *eee,* Head with corollas all strap-shaped. Single flower, left. Examples: Hawkweed, dandelion.

sepals, next the corolla made up of petals, then the stamens (pollen bearing) and finally at the center the carpels (seed bearing), which if united we may call the pistil. The two outer sets, calyx and corolla, may be spoken of together as the perianth. In some flowers only one set of parts composes the perianth in which case the set is regarded as the calyx. In our keys, since this single set is often colored and corolla-like, we have regarded the parts of such a set as petals so these

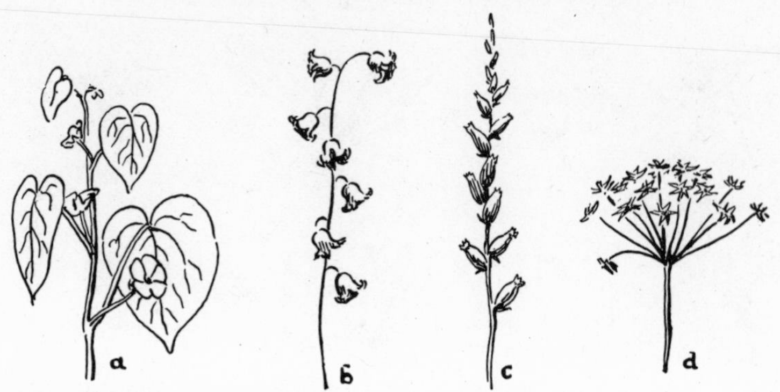

FIG. 125. TYPES OF INFLORESCENCE: A, FLOWERS SOLITARY, 1 TO A LEAF AXIL. B, FLOWERS IN A RACEME. C, FLOWERS IN A SPIKE. D, FLOWERS IN AN UMBEL.

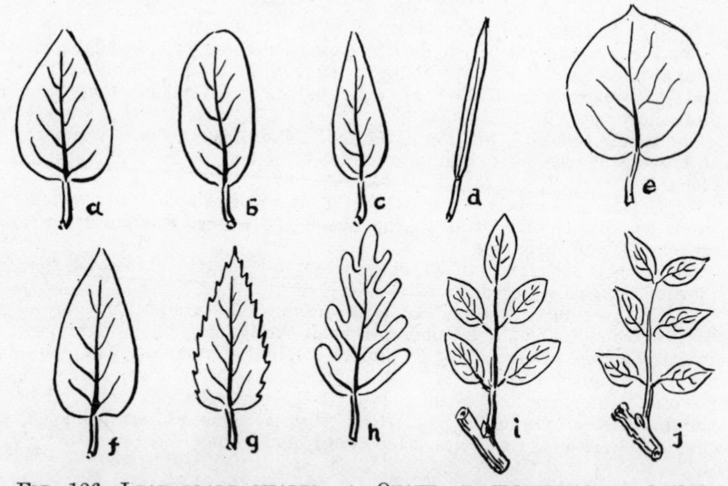

FIG. 126. LEAF BLADE SHAPES: A, OVATE; B, ELLIPTICAL; C, LANCE-SHAPED; D, LINEAR; E, ORBICULAR.
LEAF BLADE MARGIN: F, ENTIRE; G, TOOTHED; H, LOBED; I, COMPOUND WITH LEAFLETS ARRANGED UNEQUALLY PINNATE; J, COMPOUND LEAF WITH LEAFLETS EQUALLY PINNATE.

flowers could be given a place under "petal number," an important character in our keys.

A further suggestion or two in relation to flower color will be desirable. Very faint pink should also be checked under

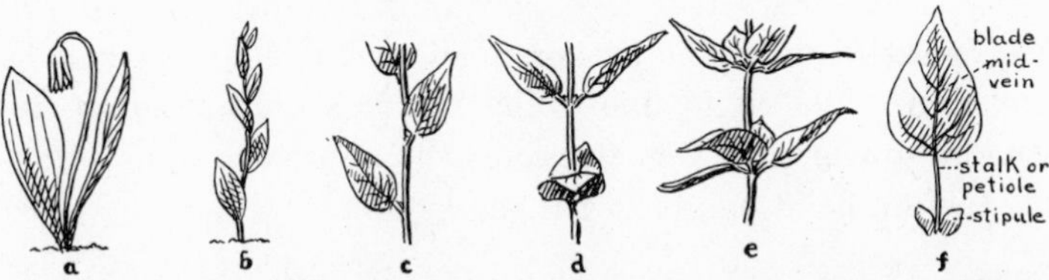

FIG. 127. LEAF ARRANGEMENT AND PARTS: A, LEAVES BASAL; B, LEAVES NOT BASAL; C, LEAVES ALTERNATE; D, OPPOSITE; E, WHORLED; F, LEAF PARTS.

"white" since many ordinarily white flowers may develop the red pigment to a slight degree under various conditions. In a flower with more than one color present, the collector should be guided by the dominant color. If the color within is different from that without the corolla, be guided by the outer hue.

In the manual or descriptive portion following the keys, the season of blooming is given. It should be kept clearly in mind that the time of the coming of spring in the east is very much earlier than that on the mountain slopes. The high mountain spring flowers are often six or even seven weeks later than those on the eastern savannahs. In reverse fashion the late asters and golden rods hold on much later in the east.

In both the key and the manual portions, all statements made refer only to those genera (indicated by number) and species which occur within the state. Since a vast majority of our species range widely into neighboring states, this book should prove useful in many adjoining areas, but its statements specifically refer to North Carolina plants and conditions.

For the satisfactory study of the smaller flowers, a hand lens is very desirable. To every gardener and amateur botanist, a high quality pocket lens will prove a continual source of delight and satisfaction, for the beauty of the tiny parts of flowers when made evident by magnification, is just as appealing as that of whole flowers.

If failure to run the keys should occur through inherent

defect in the characters used, omissions, or other error, it will be a great favor to the author if all such deficiencies will be promptly reported to him. The reader should again be reminded to learn the few characters illustrated in figures 124-127, before he attempts to use the keys.

HERBS OF WELL DRAINED SOILS

I. GREAT FOREST UPLANDS. Soil of medium to fine texture; here are included the extensive piedmont and mountian uplands whether forested or not together with those moister slopes of finer sands in the coastal plain: typical common trees, if trees are present, are the white oak, hickory, or sometimes beech. If the original forest has been destroyed, pines or scrub oak (black-jack and post) may be dominant for many years. About two-thirds of our wild flowers are included here, hence this key constitutes the main one of the book. Of the weeds, only those with showy flowers are given a place. For the Great Forest wild flowers see page 218.

II. SANDHILLS. Soil very sandy containing particles as large as granulated sugar or larger; local surface areas of glistening sand; soil thin, dark: typical common trees are the long leaf pine and turkey oak. In coastal plain. Sandhill wild flowers. See page 255.

III. COASTAL DUNES. See page 261.

HERBS OF POORLY DRAINED OR WATER COVERED AREAS

IV. SAVANNAHS. Herbs of flat, grassy or sometimes partly shrubby areas between streams or swamps; soil poorly drained in rainy periods, never permanently covered with free surface water. Pines (mostly pocosin and long leaf) may be scattered over the areas but the commonest plants are certain grasses and sedges. Chiefly in lower coastal plain. See page 262.

V. FRESH WATER MARSHES. Surface water permanent, shallow: stems or leaves projecting above the water surface: common plants cattails and wild rice. See page 269.

VI. SALT MARSHES. Salty surface water essentially permanent despite tidal fluctuation: common plants, salt grass, and salt rush. Only along our sounds and estuaries. See page 277.

VII. AQUATIC PLANTS. Surface water shallow or deep: plants rooted or free, submersed, or floating but they do not project their stems or leaves above the surface of the water: common forms, water lilies and bladder-worts. See page 279.

I. KEY TO THE GREAT FOREST WILD FLOWERS

Note: If the plant cannot be satisfactorily placed with the aid of this key, try those of the other six which from the habitat description might possibly have included the species in hand.

Plants non-green: leaves reduced to scales; stems, and leaf scales, brown, yellow or white, never green. Parasitic and saprophytic herbs. See this page below.

Plants with green leaves which are sometimes absent at time of flowering.

Flowers white or cream color. See page 219.

Flowers purple, pink or red. See page 231.

Flowers yellow or orange. See page 241.

Flowers blue, never definitely purple. See page 248.

Flowers green or green slightly modified by other colors. See page 252.

NON-GREEN HERBS: PARASITES AND SAPROPHYTES

A. Plants twining. Dodder *(Cuscuta)*, 336.

A. Plants not twining.

1. Stamens 1, united with style: orchids.

B. Lower petal (lip) under ½ in. long. Coral-root *(Corallorhiza)*, 95.

B. Lower petal (lip) over ½ in. long. Crested coral-root *(Hexalectris)*, 98.

1. Stamens 6 or more.

C. Petals united. Carolina beech drops *(Monotropsis)*, 300.

C. Petals separate.

2. Flowers solitary on end of stem. Indian pipe *(Monotropa)*, 298.

2. Flowers several. Pine-sap *(Hypopitys)*, 299.

1. Stamens 4.
   D. Flowers bilateral. Squaw-root *(Conopholis)*, 412.
   D. Flowers radial.
      3. Calyx divided into 2 lateral lobes.
         Broom-rape *(Orobanche)*, 413.
      3. Calyx equally 5-lobed.
         E. Stems short, seldom over 5 in. high.
            Naked broom-rape *(Thalesia)*, 414.
         E. Stems longer, over 5 in. high.
            Beech drops *(Leptamnium)*, 415.

Green-Leaved Herbs: Not Parasites or Saprophytes

Flowers White or Cream Color

A. Flowers in heads.
   1. Petals of small individual flowers, separate.
      Clovers *(Trifolium)*, 208.
   1. Petals united.
      B. All corollas tubular; flowers all alike.
         2. Corollas 2-lipped, bilateral, stems square. See key to the mint family, page 401.
         2. Corollas not 2-lipped; radial, stems round.
            C. Leaves alternate.
               3. Plants smooth.
                  D. Bracts of head 5. Indian plantain *(Mesadenia)*, 483.
                  D. Bracts of head more than 5.
                     4. With chaffy scales between the flowers.
                        Loud-speakers *(Marshallia)*, 475.
                     4. Without chaffy scales between the flowers.
                        E. Lower leaf petioles winged.
                           Composite fireweed *(Erechtites)*, 482.
                        E. Lower leaf petioles not winged.
                           Sweet scented Indian plantain *(Synosma)*, 484.
               3. Plants not smooth; more or less hairy.
                  F. Leaf-blades nearly smooth.
                     False boneset *(Kuhnia)*, 439.
                  F. Leaf-blades with covering of wool, at least on under side.
                     5. Basal leaves broad, 3-ribbed.
                        Everlasting *(Antennaria)*, 455.
                     5. Basal leaves narrow, not 3-ribbed.
                        G. Leaves woolly, both sides.
                           Pearly everlasting *(Anaphalis)*, 456.

G. Leaves woolly only on under side.
Cudweed *(Gnaphalium)*, 457.

C. Leaves opposite or whorled.
6. Low plants under 1 foot.
Whitlow-wort *(Paronychia)*, 108.

6. Tall plants over 1 foot.
Thoroughworts *(Eupatorium)*, 435.

B. Central flowers with tubular corollas, marginal flowers strap-shaped.
7. Leaves opposite. Leaf-cup *(Polymnia)*, 458.
7. Leaves alternate.
H. Stems winged. Crown-beard *(Verbesina)*, 472.
H. Stems not winged.
8. Marginal ray flowers few, 4-8.
I. Leaves very finely cut into numerous segments.
Yarrow *(Achillea)*, 479.

I. Leaves simple, not divided.
9. Inflorescence narrow, heads with short stalks.
White goldenrod *(Solidago)*, 434.

9. Inflorescence as broad as long, somewhat flat-topped.
J. Ray flowers over ⅛ in. long.
White-topped aster *(Sericocarpus)*, 447.

J. Ray flowers under ⅛ in. long.
Fever-few *(Parthenium)*, 462.

8. Marginal ray flowers more than 8.
K. Bracts of head in many series, the smallest at base.
10. Central flowers without chaffy scales between them.
L. Calyx a single series of fine bristles.
Aster *(Aster)*, 448.

L. Calyx absent; no fine bristles borne on top of seeds.
Field daisy *(Chrysanthemum)*, 480.

10. Central flowers with chaffy scales between them.
Northern dog-fennel *(Anthemis)*, 478.

K. Bracts of head in but ¼ series.
11. Ray flowers over ⅛ in. long.
M. Rays over 25. Fleabane *(Erigeron)*, 449.
M. Rays under 25.
12. Bushy branches; calyx of a few short scales and 2-4 bristles. Bolton's aster *(Boltonia)*, 446.
12. Not bushy branched; calyx of 2 series, outer of short scales, inner of many bristles.
False flat-topped aster *(Doellingeria)*, 451.

11. Ray flowers less than ⅛ in. long.
N. Leaves very numerous, alternate.
Horseweed *(Leptilon)*, 450.
N. Leaves not numerous, opposite.
Yerba de tajo *(Eclipta)*, 465.
B. Corollas all strap-shaped.
13. All heads hanging downward. White lettuce *(Nabalus)*, 488.
13. Heads not handing downward. Wild lettuce *(Lactuca)*, 491.
A. Flowers not in heads.
14. Petals 2.
O. Sepals green; petals notched.
Enchanters' nightshade *(Circaea)*, 272.
O. Sepals white; petals not notched.
Two-leaved Solomon's seal *(Unifolium)*, 54.
14. Petals 3, sepals often of same number, shape or color as petals (lily family and others).
P. Flowers bilateral.
15. Lower petal (lip) sac-like, inflated.
Q. Lip more than ¼ in. long.
Showy ladies' slipper *(Cypripedium)*, 76.
Q. Lip less than ¼ in. long.
Rattlesnake plantain *(Peramium)*, 92.
15. Lower petal (lip) not sac-like.
R. Inflorescence stems spirally twisted in region of the flowers.
Ladies' tresses *(Ibidium)*, 89.
R. Inflorescence stems not twisted.
16. Stems watery, purplish or brownish.
Coral-root *(Corallorhiza)*, 95.
16. Stems not watery, green in color.
S. Leaves all basal.
17. Leaves 1, linear, lance-shaped.
Pale grass pink *(Limodorum)*, 96.
17. Leaves 2, broadly elliptical in outline.
Showy orchis *(Galeorchis)*, 78.
S. Leaves not all basal, few leaves on stem.
18. Basal leaves 2, flat on ground, scale leaves on stem.
Large round-leaved orchis *(Lysias)*, 80.
18. No basal leaves on ground, stem leaves only.
T. Lower lip of flowers fringed.
White fringed orchid *(Blephariglottis)*, 81.
T. Lower lip of flowers not fringed.
Savannah white orchid *(Gymnadeniopsis)*, 84.

P. Flowers radial.

19. Leaves all basal.

U. Flowers solitary at end of stalk.

20. Flowers over 1 in. long. Atamasco lily *(Atamosco)*, 68.

20. Flowers under 1 in.

White dog-tooth violet *(Erythronium)*, 50.

U. Flowers in clusters.

21. Flowers in umbels.

V. Leaves broad, not grass-like.

Clinton's lily *(Clintonia)*, 52.

V. Leaves narrow, grass-like or tubular.

22. Plants with onion-like odor.

W. Leaves attached at stem base; mostly present at time of flowering.

Common wild onion *(Allium)*, 45.

W. Leaves not attached at flowering stem base and absent at time of flowering.

Mountain wild onion *(Validallium)*, 45.

22. Plants without onion-like odor.

False garlic *(Nothoscordum)*, 46.

21. Flowers in racemes.

X. Flowering stems over 16 in. long.

23. Corollas over 1 in. long. Bear grass *(Yucca)*, 61.

23. Corollas less than 1 in. long.

Wild hyacinth *(Quamasia)*, 51.

X. Flowering stems under 16 in. long.

Lily-of-the-valley *(Convallaria)*, 60.

19. Leaves not all basal.

Y. Leaves 3 in whorl at top of stem. Trillium *(Trillium)*, 63.

Y. Leaves scattered on stem.

24. Flowers solitary at end of stem.

Golden seal *(Hydrastis)*, 129.

24. Flowers scattered along stem in the leaf axils.

Solomon's seal *(Salomonia)*, 59.

24. Flowers in cluster at end of stem.

Z. Styles united.

25. Petals and sepals united, forming tubular corolla.

Colic root *(Aletris)*, 64.

25. Petals and sepals separate.

Wild spikenard *(Vagnera)*, 53.

Z. Styles distinct.

26. Inflorescence compound, racemes branched.

a. Sepals and petals sharp pointed at tip.
Narrow-petaled bunch flower *(Stenanthium)*, 41.
a. Sepals and petals rounded at tip.
Branching fly-poison *(Oceanoros)*, 42.
26. Inflorescence simple, one terminal raceme only.
b. Sepals and petals contracted below into a stalk.
Tall hairy bunch flower *(Melanthium)*, 44.
b. Sepals and petals not contracted below.
27. Sepals and petals 5-7 nerved.
Turkey-beard *(Xerophyllum)*, 36.
27. Sepals and petals 1 nerved.
c. Individual flower stalks less than ⅜ in. long.
Bunch-flower blazing star *(Chamaelirium)*, 37.
c. Individual flower stalks more than ⅜ in. long.
Fly poison *(Chrosperma)*, 38.
14. Petals 4.
d. Petals separate.
28. Leaves compound.
e. Two outer petals spurred or sac-like at base.
29. Plant a tender vine. Climbing fumitory *(Adlumia)*, 158.
29. Plant a low herb, not a vine.
Dutchman's breeches *(Bicuculla)*, 159.
e. No petals spurred.
30. Seeds with winged margins. Rock cress *(Arabis)*, 169.
30. Seeds without winged margins.
Bitter cress *(Cardamine)*, 168.
28. Leaves simple.
f. Leaf-blades pinnately cleft or divided.
Rock cress *(Arabis)*, 169.
f. Leaf-blades palmately cleft or divided.
Toothwort *(Dentaria)*, 167.
f. Leaf-blades not cleft or divided.
31. Flowers with ovary exserted on stalk.
Spurges *(Chamaesyce* and *Tithymalopsis)*, 242.
31. Flowers with ovary not exserted on stalk.
g. Petals and sepals borne above the ovary.
32. Flowers solitary in upper leaf axils.
Willow-herb *(Epilobium)*, 269.
32. Flowers in racemes or spikes.
Wild-herb honeysuckle *(Gaura)*, 271.
g. Petals and sepals borne below the ovary.
33. Stamens 2.
Two-leaved Solomon's seal *(Unifolium)*, 54.

33. Stamens 6.
 h. Low herbs mostly under 6 in. in height. Whitlow-grass *(Draba)*, 166.
 h. Taller herbs mostly over 6 in. in height.
  34. Leaf-blades short, nearly round in outline. Bitter cress *(Cardamine)*, 168.
  34. Leaf-blades elongated, not of round type. Rock cress *(Arabis)*, 169.
33. Stamens 4, 5 or 10. Pearlwort *(Sagina)*, 118.
33. Stamens numerous, over 10. Plants prickly. Mexican poppy *(Argemone)*, 157.

d. Petals united.
 35. Sepals and petals borne below the ovary.
  i. Calyx lobes 2. Pennywort *(Obolaria)*, 322.
  i. Calyx lobes 4.
   36. Flowers in dense terminal slender-conical spikes. Culver's root *(Leptandra)*, 396.
   36. Flowers solitary or in leafy racemes or spikes. Speedwell *(Veronica)*, 395.
  i. Calyx lobes typically 5, stems square. See key to the mints, page 401.
 35. Sepals and petals borne above the ovary.
  j. Calyx tube prolonged above ovary.
   37. Petals unequal. Wild herb honeysuckle *(Gaura)*, 271.
   37. Petals all alike. Willow herb *(Epilobium)*, 269.
  j. Calyx tube not prolonged, sepals rest directly on ovary.
   38. Plants prostrate or creeping.
    k. Stems not angled.
     39. Leaves nearly circular in outline. Partridge berry *(Mitchella)*, 420.
     39. Leaves linear or ovate, not circular in outline.
      l. Flowers solitary or but 2 or 3, clustered in leaf axils. Button-weed *(Diodia)*, 421.
      l. Flowers many (over 5) in axillary clusters. Smooth button-weed *(Spermacoce)*, 422.
    k. Stems 4-angled. Bedstraw *(Galium)*, 423.
   38. Plants erect.
    m. Stems 4-angled. Bedstraw *(Galium)*, 423.
    m. Stems not 4-angled.
     40. Corollas funnel form. Bluet *(Houstonia)*, 419.
     40. Corollas wide-spreading, tube short. Clustered bluets *(Oldenlandia)*, 418.

14. Petals 5 or more (sepals to be regarded as petals, if petals are not not present).
  n. Petals separate.
    41. Flowers radial.
      o. Leaves chiefly basal.
        42. Flowers solitary at end of stalk.
          p. Leaves simple.
            43. Plants with reddish juice; leaves 5-9 lobed. Blood-root *(Sanguinaria)*, 156.
            43. Plants without reddish juice; leaves 3-lobed. Gold-thread *(Coptis)*, 131.
          p. Leaves compound.
            44. Leaflets 3, entire, inverted heart-shaped. Mountain wood sorrel *(Oxalis)*, 235.
            44. Leaflets 3 or 5, much toothed and cut. Wind-flower *(Anemone)*, 139.
        42. Flowers in clusters.
          q. Plants with milky juice. Spurge *(Tithymalus)*, 242.
          q. Plants without milky juice.
            45. Carpels many, separate.
              r. Leaves simple, palmately lobed.
                46. Stamens 5. Alum- root *(Heuchera)*, 184.
                46. Stamens more than 5. *False bugbane (Trautvetteria)*, 143.
              r. Leaves compound.
                47. Leaflets 3. Wild Strawberry *(Fragaria)*, 193.
                47. Leaflets more than 3.
                  s. Leaflets sessile. Wind-flower *(Anemone)*, 139.
                  s. Leaflets stalked. Rue anemone *(Syndesmon)*, 141.
            45. Carpels united into a single pistil, or only 1 carpel present.
              t. Flowers less than ½ in. dia.
                48. Ovary 1-celled Foam-flower *(Tiarella)*, 185.
                48. Ovary 2-celled. Saxifrage *(Micranthes)*, 187.
              t. Flowers over ½ in. dia.
                49. Leaves large, over 4 in. across, lobed. Twin-leaf *(Jeffersonia)*, 153.
                49. Leaves smaller, under 4 in., not lobed. False wintergreen *(Pyrola)*, 296.
      o. Leaves not chiefly basal.
        50. Leaves 3 or more in whorls.

u. Leaves 6 or less in 1 whorl at base of terminal flower stalk. Rue anemone *(Syndesmon)*, 141.

u. Leaves numerous, more than 6 in many whorls on stem. Spurry *(Spergula)*, 117.

50. Stem leaves opposite.

v. Only 2 leaves on flowering stem above ground.

51. Leaves not over 6 in. long.

w. Sepals 2. Spring beauty *(Claytonia)*, 114.

w. Sepals 5. Bishop's cap *(Mitella)*, 183.

51. Leaves over 6 in. long.

x. Flower solitary between the two leaves. May-apple *(Podophyllum)*, 152.

x. Flowers in compound cluster. Umbrella-leaf *(Diphylleia)*, 154.

More than 2 leaves on stems.

52. Flowers with ovary exserted on stalk. Spurges *(Chamaesyce* and others), 242.

52. Flowers with ovary sessile, not exserted.

y. Petals notched or cleft at tip, sometimes nearly to the base.

53. Sepals united.

z. Calyx tube 5-ribbed. Bouncing Bet *(Saponaria)*, 125.

z. Calyx tube 10-ribbed. Wild pinks *(Silene)*, 126.

53. Sepals separate.

aa. Styles 3-4. Chickweed *(Alsine)*, 121.

aa. Styles 5. Mouse-ear chickweed *(Cerastium)*, 122.

y. Petals not cleft at tip.

54. Stamens 3 or 5.

bb. Both calyx and corolla present. Jagged chickweed *(Holosteum)*, 119.

bb. Calyx apparently absent, only one set of perianth parts. Nailwort *(Paronychia)*, 108.

54. Stamens 10.

cc. Styles as many as the sepals. Pearlwort *(Sagina)*, 118.

cc. Styles fewer than the sepals. Wild chickweed *(Alsinopsis)*, 120.

50. Stem leaves alternate.

dd. Flowers small, strictly in umbels.

55. Tall plants, over 4 ft. high.
  ee. Stems smooth below; leaves under 2 ft. wide.
  Angelica *(Angelica)*, 293.
  ee. Stems hairy below; compound leaf often over 2 ft. wide. Cow-parsnip *(Heracleum)*, 295.
55. Medium or low plants, under 4 ft. high.
  ff. Plants low, not over 9 in. high.
  Harbinger-of-spring *(Erigenia)*, 281.
  ff. Plants medium, over 9 in. high.
    56. Leaves palmately divided.
      gg. Leaf divisions 3; no leaf-like bracts at base of inflorescence. Honewort *(Deringa)*, 286.
      gg. Leaf divisions 5; leaf-like bracts present at base of inflorescence.
      Snake-root *(Sanicula)*, 277.
    56. Leaves pinnately divided.
      hh. Fruiting bodies slender, at least ½ in. long.
      Sweet cicely *(Washingtonia)*, 280.
      hh. Fruiting bodies ovoid, ¼ in. long or less.
        57. Surface of fruit smooth.
          ii. Leaflets ovate or nearly so, not cut into fine segments. Nondo *(Ligusticum)*, 291.
          ii. Leaflets deeply cut into rather fine segments. Chervil *(Chaerophyllum)*, 283.
        57. Surface of fruit rough or bristly.
          jj. Flowers of ultimate umbels, over 10.
          Queen Anne's lace *(Daucus)*, 279.
          jj. Flowers of ultimate umbels, not over 10.
          Scaly seed *(Spermolepis)*, 282.

dd. Flowers not in umbels.
58. Leaves simple.
  kk. Leaves fleshy, mostly cylindric in shape, low plant.
    59. Carpels 4.
    Four-fruited stonecrop *(Tetrorum)*, 178.
    59. Carpels 5. Stonecrop *(Sedum)*, 179.
  kk. Leaves not fleshy, plants not low or spreading.
    60. Leaf-blades with 5 or more prominent lobes.
      ll. Plants with stinging hairs.
      Tread softly *(Cnidoscolus)*, 241.
      ll. Plants without stinging hairs.
      Aconite saxifrage *(Therophon)*, 186.
    60. Leaf-blades not lobed.

mm. Leaf-blades entire, at least below the middle. Pokeweed *(Phytolacca)*, 110.

mm. Leaf-blades toothed.

61. Leaf-blades coarsely toothed above middle. Mountain spurge *(Pachysandra)*, 244.

61. Leaf-blades with few spreading teeth along entire sides, white veined. Spotted wintergreen *(Chimaphila)*, 297.

7. 58. Leaves compound.

nn. Leaflets 3.

62. Plants erect. Indian physic *(Porteranthus)*, 191.

62. Plants trailing.

oo. On high mountain tops only. Three-toothed cinquefoil *(Sibbaldiopsis)*, 192.

oo. Not on high mountain tops. Wild strawberry *(Fragaria)*, 193.

nn. Leaflets more than 3.

63. Inflorescence a simple raceme.

pp. Flowers prominently spurred. Larkspur *(Delphinium)*, 136.

pp. Flowers not spurred.

64. Racemes not over 6 in. in length. Baneberry *(Actaea)*, 134.

64. Racemes over 6 in. in length. Black cohosh *(Cimicifuga)*, 133.

63. Inflorescence other than a simple raceme.

qq. Styles crooked, bayonet-shaped. Avens *(Geum)*, 197.

qq. Styles not crooked, sometimes curved.

65. Inflorescence under 4 in. in length. False rue-anemone *(Isopyrum)*, 132.

65. Inflorescence over 4 in. in length.

rr. Flower stalks more than 1/16 in. long. Goat's-beard *(Aruncus)*, 190.

rr. Flower stalks less than 1/16 in. long. False goat's beard *(Astilbe)*, 189.

41. Flowers bilateral.

ss. Leaves simple, sometimes deeply lobed.

66. Calyx apparently absent, only colored "petals" present. Monk's hood *(Aconitum)*, 137.

66. Calyx present.

tt. Upper 3 petals all alike, narrowed at base, no petals spurred. High mountain saxifrage *(Hydatica)*, 188.

tt. Upper 2 petals alike, lower one largest and slightly spurred. White violets *(Viola)*, 254.

ss. Leaves compound.

67. Plants twining.

uu. Leaves terminated by tendrils. Vetch *(Vicia)*, 230.

uu. Leaves not terminated by tendrils, but by leaflets. Hog-peanut *(Falcata)*, 225.

67. Plants not twining.

vv. Leaves much divided into numerous fine leaflets. Dutchman's breeches *(Bicuculla)*, 159.

vv. Leaves with 3 leaflets.

68. Stamen stalks united, forming a tube. Tick-trefoil *(Meibomia)*, 220.

68. Stamen stalks separate, not united into a tube. White wild indigo *(Baptisia)*, 205.

n. Petals united.

69. Flowers radial.

ww. Sepals and petals borne above ovary.

70. Leaves opposite. Corn salad *(Valerianella)*, 425.

70. Leaves alternate.

xx. Petals short on rim of bell-shaped corolla. Bellflower *(Campanula)*. 428.

xx. Petals lance-shaped, spreading from short corolla tube. American bellflower *(Campanulastrum)*, 429.

ww. Sepals and petals borne below ovary.

71. Plants trailing and twining.

yy. Flowers over 1½ in. broad. Wild morning-glory *(Ipomoea* and *Convolvulus)*, 334.

yy. Flowers less than 1½ in. broad.

72. Plants creeping on ground. Trailing arbutus *(Epigaea)*, 301.

72. Plants vine-like. Sand vine *(Gonolobus)*, 329.

71. Plants erect, not trailing or twining.

aaa. Leaves basal.

73. Flowers solitary at ends of stalks.

bbb. Leaves nearly circular in outline. Shortia *(Shortia)*, 305.

bbb. Leaves definitely longer than broad, not circular. American cowslip *(Dodecatheon)*, 312.

73. Flowers in a slender raceme. Galax *(Galax)*, 306.

aaa. Leaves not basal.

74. Leaves simple.
ccc. Flowers in an umbel.
75. Flowers many, over 10 in cluster.
Milkweed *(Asclepias)*. 328.
75. Flowers under 10 in cluster.
Night shade and horse nettle *(Solanum)*, 345.
ccc. Flowers in a raceme or spike-like inflorescence.
76. Flowers over ¼ in. dia.
Wild comfrey *(Cynoglossum)*, 347.
76. Flowers less than ¼ in. dia.
ddd. Flowers with bracts beneath them.
77. Fruits smooth. Stickseed *(Lappula)*, 348.
77. Fruits spiny.
Gromwell *(Lithospermum)*, 351.
ddd. Flowers without bracts evident.
78. Corolla tube twice as long as the calyx.
Heliotrope *(Heliotropium)*, 353.
78. Corolla tube little longer than the calyx.
For-get-me-not *(Myosotis)*, 350.
ccc. Flowers not in racemes or umbels.
79. Stems not spiny. Wild phlox *(Phlox)*, 342.
79. Stems spiny. Horse nettle *(Solanum)*, 345.
74. Leaves compound.
eee. Stamens exserted from corolla tube.
Water-leaf *(Hydrophyllum)*, 337.
eee. Stamens not exserted.
80. Appendages in notches of calyx tube.
Creeping water-leaf *(Nemophila)*, 338.
80. No appendages in notches of calyx tube.
Small leaved water-leaf *(Phacelia)*. 340.
59. Flowers bilateral.
fff. Stems square.
81. Anther bearing stamens 4.
ggg. Corollas not 2-lipped or slightly so.
Mint *(Mentha)*, 379.
ggg. Corollas definitely 2-lipped.
82. Leaf-blade base continuing down sides of leaf stalk.
Bitter mint *(Mesosphaerum)*, 382.
82. Leaf-blade base not continuing down stalk.
hhh. Leaves sharply toothed.
Hedge hyssop *(Mecardonia)*, 391.
hhh. Leaves wavy-margined.
Basil *(Clinopodium)*, 374.

81. Anther bearing stamens 2.
  iii. Corolla tubes not 2-lipped, petals about equal.
  Water hoarhound *(Lycopus)*, 378.
  iii. Corolla tubes 2-lipped.
    83. Stamens exserted. Stone mint *(Cunila)*, 377.
    83. Stamens included.
      jjj. Calyx 2-lipped. Sage *(Salvia)*, 370.
      jjj. Calyx not 2-lipped. Horse mint *(Monarda)*, 371.

fff. Stems not square.
  84. Ovary 1-celled.
    kkk. Flowers large over ½ in. long, fruit with 2 large hooks. Unicorn-plant *(Martynia)*, 416.
    kkk. Flowers small, not over ½ in. long, fruit not hooked. Lop seed *(Phryma)*, 356.
  84. Ovary 2-celled, 2 carpels united.
    lll. Flowers with 4 normal stamens and one sterile one, smooth or bearded.
      85. Sterile stamen smooth. Turtle-head *(Chelone)*, 386.
      85. Sterile stamen bearded.
      Beard-tongue *(Penstemon)*, 387.
    lll. Flowers with 2 or 4 normal stamens, no sterile one.
      86. Stamens 4.
        mmm. Leaf margins cut into numerous segments.
        Lousewort *(Pedicularis)*, 404.
        mmm. Leaf margins entire.
        Cow-wheat *(Melampyrum)*, 403.
      86. Stamens 2. Culver's root *(Leptandra)*, 396.
    lll. Flowers with 5 stamens, all exserted from cleft on upper side. Lobelia *(Lobelia)*, 431.

Flowers Purple, Pink, or Red

A. Flowers in heads.
  1. Petals of small individual flowers separate.
  Clovers *(Trifolium)*, 208.
  1. Petals united.
    B. Corollas all tubular; bilateral, 2-lipped: stems square.
    See key to the mints, page 401.
    B. Corollas all tubular; radial, stems mostly round.
      2. Leaves chiefly basal.
        C. Clusters of heads surrounded by 2-3 prominent leaf-like bracts. Elephant's foot *(Elephantopus)*, 433.
        C. Heads not surrounded by broad leafy bracts.

3. Heads solitary on end of stalk.
Loud-speakers *(Marshallia)*, 475.
3. Heads many on stalk, arranged in flat-topped cluster.
False blazing star *(Carphephorus)*, 438.
2. Leaves not chiefly basal.
D. Leaves pinnate. Sensitive brier *(Morongia)*. 201.
D. Leaves simple.
4. Leaves whorled, 3-6 at one node.
Joe-Pye weed *(Eupatorium)*, 435.
4. Leaves alternate.
E. Leaves very spiny. Thistle *(Carduus)*, 486.
E. Leaves not spiny.
5. With chaffy scales between the flowers.
Loud-speakers *(Marshallia)*, 475.
5. Without chaffy scales between the flowers.
F. Leaf-blades generally narrow (linear), entire.
Blazing star *(Laciniaria)*, 437.
F. Leaf-blades broad, toothed.
Ironweed *(Vernonia)*, 432.
4. Leaves opposite. Mountain mint *(Koellia)*, 375.
B. Central corollas tubular, marginal ray corollas strap-shaped.
6. Ray flowers numerous, over 40.
Robin's plantain *(Erigeron)*, 449.
6. Ray flowers under 40.
G. Leaves opposite. Rose colored tickseed *(Coreopsis)*, 473.
G. Leaves alternate.
7. Rays over 1 in. long.
Purple cone-flower *(Brauneria)*, 468.
7. Rays under 1 in. long. Aster *(Aster)*, 448.
B. Corollas all strap-shaped; bilateral, not 2-lipped.
Salsify *(Tragopogon)*, 493.
A. Flowers not in heads.
8. Peculiar small bilateral flowers which have a slender tubular corolla of 3 united petals projecting between 2 colored or petal-like sepals; stamens 8. Milkwort *(Polygala)*, 238.
8. Flowers not as above.
H. Petals 2, upper with 2 or 3 small teeth, lower entire.
False mint *(Diapedium)*, 408.
H. Petals 3. Sepals often of same number, shape or color as petals (lily family and others). Sepals to be regarded as petals if petals are not present.
9. Flowers radial.
I. Leaves chiefly basal.

10. Flowers in an umbel on tall stem. Wild onion *(Allium)*, 45.
10. Flowers 1-3 on short stems, very close to the ground under leaves.
J. Styles united. Wild ginger *(Asarum)*, 427.
J. Styles distinct. Wild ginger *(Hexastylis)*, 427.
I. Leaves not chiefly basal.
11. Leaves 3 in whorl at summit of stem. Trillium *(Trillium)*, 63.
11. Leaves many, scattered on stem.
K. Flowers over 1 in. long. Lily *(Lilium)*, 48.
K. Flowers under 1 in. long.
12. Flowers solitary or in pairs with no bracts. Twisted stalk *(Streptopus)*, 55.
12. Flowers in clusters partly surrounded by a prominent leaf-like bract. Spiderwort *(Tradescantia)*, 28.
9. Flowers bilateral.
L. Sepals 5, two larger than the rest. Milkwort *(Polygala)*, 238.
L. Sepals 3, (orchids).
13. Stem leaves scale-like, at least at base of stem.
M. Scales at top of stem leaf-like, whorled. Whorled snake-mouth *(Isotria)*, 87.
M. Scales at top of stem not leaf-like, very small.
14. Lower petals (lips) not lobed, merely notched at apex. Coral-root *(Corallorhiza)*, 95.
14. Lower petals (lips) prominently 2-lobed. Twayblade *(Ophrys)*, 90.
14. Lower petals (lips) 3-lobed. Crested coral-root *(Hexalectris)*, 98.
13. Stem leaves not scale-like.
N. Leaves all basal.
15. Leaves oval, broad.
O. Flower base enclosed by leaf-like structure. Showy orchis *(Galeorchis)*, 78.
O. Flower base not enclosed by leaf-like structure. Large twayblade *(Liparis)*, 94.
15. Leaves narrow. Bog arethusa *(Arethusa)*, 88.
N. Leaves not basal.
16. Not more than 2 leaves on stems. Snake-mouth orchis *(Pogonia)*, 85.
16. More than 2 leaves on stems.
P. Lower petal (lip) sac-like. Showy ladies' slipper *(Cypripedium)*, 76.

P. Lower petal (lip) flattened, not sac-like.
17. Flowers nodding.
Nodding pogonia *(Triphora)*, 86
17. Flowers not nodding.
Fringed orchids *(Blephariglottis)*, 81.

H. Petals 4.
18. Flowers radial.
Q. Sepals and petals borne above the ovary.
19. Upper leaves strictly opposite.
R. Calyx contracted above the ovary into a short tube.
Meadow beauty *(Rhexia)*, 258.
R. Calyx not contracted above the ovary into a tube.
Bluet *(Houstonia)*, 419.
19. Upper leaves alternate.
S. Calyx contracted above the ovary into a slender tube.
Willow-herb *(Epilobium)*, 269.
S. Calyx not contracted, sepals arising directly from ovary.
Fireweed *(Chamaenerion)*, 268.
Q. Sepals and petals borne below the ovary.
20. Petals united. Pennywort *(Obolaria)*, 322.
20. Petals separate.
T. Leaves not over ¼ in. long, fleshy.
Southern rock orpine *(Diamorpha)*, 176.
T. Leaves over ¼ in. long, not fleshy.
21. Leaves simple; stamens not 2 or 3 times as long as the petals.
U. Petals narrow, under ⅜ in. wide.
Toothwort *(Dentaria)*, 167.
U. Petals broad, over ⅜ in. wide.
Meadow-beauty *(Rhexia)*, 258.
21. Leaves palmately compound; stamens 2-3 times as long as the petals. Spider-flower *(Cleome)*, 171.
18. Flowers bilateral.
V. Petals united.
22. Stems square. (See key to the mints), page 401.
22. Stems round. False mint *(Diapedium)*, 408.
V. Petals separate.
23. Two outer petals spurred or sac-like at base.
W. Plants tender vines. Allegheny-vine *(Adlumia)*, 158.
W. Plants not vines. Wild bleeding heart *(Bicuculla)*, 159.
23. One outer petal only, spurred.
Roman wormwood *(Capnoides)*, 160.

H. Petals 5 or more. Sepals to be regarded as petals, if petals are not present.

24. Petals separate.

X. Flowers radial.

25. Leaves compound.

Y. Plants over 8 in.

26. Petals long-spurred. Columbine *(Aquilegia)*, 135.

26. Petals not spurred.

Z. Flowers in erect umbels.
Meadow parsnip *(Thaspium)*, 292.

Z. Flowers in loose clusters, drooping.
Meadow-rue *(Thalictrum)*, 146.

Y. Plants under 8 in. tall.

27. Underground stems bulb-like; petals uniformly colored. Violet woodsorrel *(Ionoxalis)*, 236.

27. Underground stems slender, branching; veins of petals more deeply colored.
Woodsorrel *(Oxalis)*, 235.

25. Leaves simple.

a. Flowers solitary.

28. Leaves lobed.

b. Plants prickly. Mexican poppy *(Argemone)*, 157.

b. Plants not prickly.

29. Twining vines, flowers over 2 in. in dia.
Passion flower *(Passiflora)*, 256.

29. Not twining, leaves all basal, flowers under 1 in. diam. Hepatica *(Hepatica)*, 138.

28. Leaves not lobed, narrow.
Corn cockle *(Agrostemma)*, 127.

a. Flowers in erect narrow racemes.

30. Stems with collar-like appendages above the leaves.

c. Leaf-blades jointed at base; stem collars 2-lobed.
Smartweed *(Polygonum)*, 102.

c. Leaf-blades not jointed at base; stem collars not 2-lobed. Smartweed *(Persicaria)*, 102

30. Stems without collar-like appendages.
Milkworts *(Polygala)*, 238.

a. Flowers in low flat-topped inflorescence.

32. Leaves chiefly basal. Alum-root *(Heuchera)*, 184.

32. Leaves not chiefly basal.

d. Leaves opposite.

33. Leaf-blades palmately lobed.
Cranesbill *(Geranium)*, 232.

33. Leaf-blades not palmately lobed.
    e. Stem leaves 2. Spring beauty *(Claytonia)*, 114.
    e. Stem leaves many.
        34. Calyx tube with 5 ribs.
            f. Petal blades with an appendage at base. Bouncing Bet *(Saponaria)*, 125.
            f. Petal blades without appendage at base.
                35. Calyx base surrounded by bracts. Pink *(Dianthus)*, 123.
                35. Calyx base not surrounded by bracts. Cow-herb *(Vaccaria)*, 124.
        34. Calyx tube with 10 ribs. Fire pink *(Silene)*, 126.

d. Leaves alternate.
    36. Leaves narrow, cylindric, fleshy; stamens not united. Fame flower *(Talinum)*, 113.
    36. Leaves flat, broad; stamens united into a tube. Poppy-mallow *(Callirrhoe)*, 247.

X. Flowers bilateral.
    37. Leaves simple.
        h. Petals 5.
            38. Flowers solitary. Violets *(Viola)*, 254.
            38. Flowers in racemes, two of the "petals" being enlarged sepals. Milkwort *(Polygala)*, 238.
        h. Petals 6. Blue wax-weed *(Parsonia)*, 263.
    37. Leaves compound.
        i. Leaflets 2. Wild sweet pea *(Lathyrus)*, 231.
        i. Leaflets 3.
            39. Plants vine-like, trailing or twining.
                j. Keel (made of lower two petals of corolla) coiled or curved.
                    40. Keel merely incurved, not spirally twisted. Wild bean *(Strophostyles)*, 229.
                    40. Keel spirally twisted. Bean vine *(Phaseolus)*, 228.
                j. Keel not curved or twisted.
                    41. Flowers large, 1 in. or more long.
                        k. Styles hairy on inner side; flowers over 1½ in. long. Butterfly pea *(Clitoria)*, 227.
                        k. Styles pubescent only at tip, flowers not over 1½ in. long. Virginia butterfly pea *(Bradburya)*, 226.

41. Flowers less than 1 in. long.
  1. Calyx with 2 bractlets at base.
   Milk pea *(Galactia)*, 224.
  1. Calyx without bractlets.
   42. Upper large petal spurred at base.
    Virginia butterfly pea *(Bradburya)*, 226.
   42. Upper large petal not spurred.
    Hog peanut *(Falcata)*, 225.

39. Plants not vine-like.
 m. Pods commonly 1-seeded.
  43. Flowers in short clusters in axils of upper reduced leaves. Bush clover *(Lespedeza)*, 221.
  43. Flowers on stalked elongated racemes.
   Blue darts *(Psoralea)*, 214.
 m. Pods with more than 1 seed.
  44. Pods prominently jointed.
   Tick-trefoil *(Meibomia)*, 220.
  44. Pods not jointed.
   Bird's-foot trefoil *(Lotus)*, 209.

i. Leaflets more than 3.
 45. Leaves not terminated by a tendril.
  Goat's-rue *(Cracca)*, 210.
 45. Leaves terminated by a tendril.
  n. Side petals of flowers attached to lower petals at base. Vetch *(Vicia)*, 230.
  n. Side petals nearly free.
   Wild sweet pea *(Lathyrus)*, 231.

24. Petals united.
 o. Flowers radial.
  46. Leaves all basal. American cow-slip *(Dodecatheon)*, 312.
  46. Leaves not all basal.
   p. Plants vine-like, trailing or twining.
    47. Stems 4-angled, prickly.
     Tear thumb *(Tracaulon)*, 104.
    47. Stems round, not prickly.
     q. Stems woody, flowers under ½ in. long.
      Trailing arbutus *(Epigaea)*, 301.
     q. Stems herbaceous, flowers over ½ in. long.
      Bindweed *(Convolvulus)*, 335.
   p. Plants not vine-like.
    48. Leaves opposite or whorled.
     r. Flowers in a spike.

49. Flowers projecting on one side of spike only.
Carolina pink *(Spigelia)*, 314.

49. Flowers projecting on all sides of spike.
Vervain *(Verbena)*, 354.

r. Flowers in close clusters in upper leaf axils.

50. Flowers not over ½ in. long.
Mint *(Mentha)*, 379.

50. Flowers over ½ in. long.

s. Sepals and petals borne above the ovary.
Horse gentian *(Triosteum)*, 424.

s. Sepals and petals borne below the ovary.

51. Petals large, spreading at end of funnel-shaped corolla tube.
Twin blue bells *(Ruellia)*, 407.

51. Petals merely teeth on margin of the corolla tube.

t. Corollas with plaits between lobes.
Gentian *(Dasystephana)*, 318.

t. Corollas without plaits between lobes.
Ague-weed *(Gentianella)*, 317.

r. Flowers in open, loose clusters on upper part of plant; not in close clusters in upper leaf axils.

52. Tube of corolla under ⅜ in. long, flaring.

u. Flowers in an umbel.
Milkweed *(Asclepias)*, 328.

u. Flowers not in an umbel.

53. Ovary 1-celled, stem not much branched below the middle. Sabbatia *(Sabbatia)*, 316.

53. Ovary 2-celled, stem branched below middle.
Dogbane *(Apocynum)*, 325.

52. Tube of corolla over ⅜ in. long, petals spreading from end of tube.

v. Corolla tube very slender. Phlox *(Phlox)*, 342.

v. Corolla tube funnel-form.

54. Leaf-blades ovate with a single slender lobe borne from base.
Eared gerardia *(Otophylla)*, 400.

54. Leaf-blades linear.
Gerardia *(Gerardia)*, 399.

48. Leaves alternate.

w. Stems prickly.

55. Prickles prominent scattered on round stems.
Horse nettle *(Solanum)*, 345.

55. Prickles fine, in rows on angles of square stem. Tear thumb *(Tracaulon)*, 104.

w. Stems not prickly.

56. Leaf-blades entire or slightly toothed.

x. Flowers solitary in the axils of sessile leaf blades. Venus' looking-glass *(Specularia)*, 430.

x. Flowers in terminal racemes.

57. Five scales in throat of corolla. Wild comfrey *(Cynoglossum)*, 347.

57. No scales in corolla throat. Twin-fruit *(Amsonia)*, 324.

56. Leaf-blades coarsely lobed or toothed.

y. Flowers over 3 in. long. Jimson-weed *(Datura)*, 346.

y. Flowers under 1 in. long.

58. Stamens included in corolla tube. Small-leaved water-leaf *(Phacelia)*, 340.

58. Stamens exserted from corolla tube. Water-leaf *(Hydrophyllum)*, 337.

o. Flowers bilateral.

59. Stems square.

z. Anther bearing stamens 2.

60. Calyx 2-lipped, lobes unequal.

aa. Leaves chiefly basal. Sage *(Salvia)*, 370.

aa. Leaves not basal.

61. Leaf-blades not over ¾ in. long. Pennyroyal *(Hedeoma)*, 373.

61. Leaf-blades over 1 in. long. False horsemint *(Blephilia)*, 372.

60. Calyx not 2-lipped, lobes equal.

bb. Corollas brilliant scarlet. Bee balm *(Monarda)*, 371.

bb. Corollas not scarlet.

62. Flowers not over ¼ in. long. Stone mint *(Cunila)*, 377.

62. Flowers over ¼ in. long. Horse mint *(Monarda)*, 371.

z. Anther bearing stamens 4.

63. Calyx definitely 2-lipped.

cc. Leaves not over ¼ in. long. Thyme *(Thymus)*, 376.

cc. Leaves over ¼ in. long.

64. Flowers in loose or open terminal cluster. Stone root *(Micheliella)*, 380.

64. Flowers in compact terminal or axillary clusters.

dd. Flowers in single, terminal, spike-like inflorescence. Self-heal *(Prunella)*, 365.

dd. Flowers in axillary clusters spaced apart. Basil *(Clinopodium)*, 374.

63. Calyx not 2-lipped.

ee. Stamens exserted from cleft in upper lip of corolla. American germander *(Teucrium)*, 357.

ee. Stamens not exserted from cleft in upper lip.

65. Leaves nearly circular in outline with rounded teeth. Henbit *(Lamium)*, 369.

65. Leaves of the ovate type usually serrate.

ff. Stamens straight.

66. Flowers mostly in terminal heads. Mountain mint *(Koellia)*, 375.

66. Flowers in scattered axillary clusters. Mint *(Mentha)*, 379.

ff. Stamens curved.

67. Upper pair of stamens longer than lower pair. Giant hyssop *(Agastache)*, 361.

67. Lower pair of stamens longer than upper pair.

gg. Upper lip of corolla flat, not concave. Bitter mint *(Mesosphaerum)*, 382.

gg. Upper lip concave.

68. Corollas about 1 in. long. Dragon-head *(Physostegia)*, 367.

68. Corollas about ⅜ in. long. Hedge nettle *(Stachys)*, 368.

59. Stems not square.

hh. Anther bearing stamens 4.

69. Corollas with bearded stamen-like structure on lower side within. Beard-tongue *(Penstemon)*, 387.

69. Corollas without such stamen-like structure.

ii. Corollas spurred. Toad flax *(Linaria)*, 384.

ii. Corollas not spurred.

70. Upper lip of corolla external in the flower bud. Turtle-head *(Chelone)*, 386.

70. Lower lip of corolla external in the flower bud.

jj. Flowers and accompanying bracts brilliant red. Painted cup *(Castilleja)*, 402.

jj. Flowers purple, bracts not colored.

hh. Anther bearing stamens 5.

71. Corollas brilliant red.

Cardinal flower *(Lobelia)*, 431.

71. Corollas purplish. Lobelia *(Lobelia)*, 431.

hh. Anther bearing stamens 8.

Milkwort *(Polygala)*, 238.

hh. Anther bearing stamens 10, filaments united.

Hairy goat's-rue *(Cracca)*, 210.

Flowers Yellow or Orange

A. Flowers in heads.

Aa. Petals separate, in the small individual flowers.

Clovers *(Trifolium)*, 208.

Aa. Petals of all flowers united.

1. No corollas strap-shaped, flowers all alike. Peculiar flowers with two sepals larger than the others, petal-like; petals united into a tube. Milkworts *(Polygala)*, 238.

1. Marginal corollas strap-shaped, central ones tubular.

B. Leaves chiefly basal.

2. Heads small, numerous, less than 1½ in. dia.

Ragworts *(Senecio)*, 485.

2. Heads large, few, over 1½ in. dia.

Leopard's-bane *(Arnica)*, 481.

B. Leaves not chiefly basal.

3. Chaffy scales not present among flowers attached to receptacle at flower base.

C. Heads over ¾ in. dia.

4. Rays reddish purple without.

Gaillardia *(Gaillardia)*, 477.

4. Rays without purple color.

D. Strap-shaped corollas, minutely lobed if at all; pappus of scales and bristles. Golden aster *(Chrysopsis)*, 440.

D. Strap-shaped corollas, prominently three lobed at tip; pappus of scales only. Sneezeweeds *(Helenium)*, 476.

C. Heads under ¾ in. dia.

5. Plants much branched above, forming flat-topped inflorescence. Flat-topped goldenrod *(Euthamia)*, 443.

5. Not forming flat-topped inflorescence.

E. Basal leaves heart-shaped, with slender petioles.

False goldenrod *(Brachychaeta)*, 445.

E. Basal leaves not definitely heart-shaped, with broad petioles. Goldenrod *(Solidago)*, 434.

3. Chaffy scales present among flowers, attached to receptacle at base of flowers.
  F. Only ray flowers seed producing. (Note size of seeds (achenes) at base of fruiting heads).
    6. Rays 5. Spring green and gold *(Chrysogonum)*, 460.
    6. Rays more than 5.
      G. Bracts in 2 series, the outer of 5 broad bracts. Leaf cup *(Polymnia)*, 458.
      G. Bracts in more than 2 series, not distinct. Rosin-weed *(Silphium)*, 459.
  F. Disk (central tubular) flowers seed producing, ray flowers sterile.
    7. Lower leaves opposite.
      H. Disk flower mass conical, longer than broad. False coneflower *(Spilanthes)*, 466.
      H. Disk flower mass conical, broader than long.
        8. Four outer bracts large, leaf-like, remainder smaller. Hairy false sunflower *(Tetragonotheca)*, 464.
        8. All bracts narrow. Smooth false sunflower *(Heliopsis)*, 463.
      H. Disk flower mass flat.
        9. Seeds with barbed teeth. Beggarticks *(Bidens)*, 474.
        9. Seeds with 2 unbarbed teeth. Tickseed *(Coreopsis)*, 473.
    7. Lower leaves alternate.
      I. Stems winged.
        10. Bracts of heads many, in 2 or more series. Crown-beard *(Verbesina)*, 472.
        10. Bracts of heads few. Wing-stem *(Actinomeris)*, 471.
      I. Stems not winged.
        11. Central flower mass flattened or merely convex. Sunflower *(Helianthus)*, 470.
        11. Central flower mass conical in outline. Coneflower *(Rudbeckia)*, 467.

1. All corollas strap-shaped; not 2-lipped.
  J. Leaves chiefly basal.
    12. Flowering stem hollow, tubular. Dandelion *(Taraxacum)*, 492.
    12. Flowering stem solid.
      K. Leaves smooth. False dandelion *(Adopogon)*, 494.
      K. Leaves hairy. Hawkweed *(Hieracium)*, 489.
  J. Leaves not chiefly basal.
    13. Under 2 feet tall. False dandelion *(Sitilias)*, 490.
    13. Over 2 ft. tall. Yellow wild lettuce *(Lactuca)*, 491.

GREAT FOREST YELLOW FLOWERS

1. All corollas tubular, 2-lipped, stems square.
(See key to the mints), page 401.

A. Flowers not in heads.

14. Petals numerous over 15, united below. Cactus *(Opuntia)*, 257.

14. Petals 3; sepals often of same number, shape or color as petals (lily family and others).

L. Leaves chiefly basal.

15. Flowers solitary. Yellow dog tooth violet *(Erythronium)*, 50.

15. Flowers in clusters.

M. Leaves narrow, grass-like.

16. Perianth borne on top of ovary.
Yellow star grass *(Hypoxis)*, 66.

16. Perianth borne below the ovary.

N. Flowers less than ½ in. dia.
Bunch-flower blazing star *(Chamaelirium)*, 37.

N. Flowers over 1 in. dia. Day lily *(Hemerocallis)*, 47.

M. Leaves broad, oval.

17. Flowers radial. Clinton's lily *(Clintonia)*, 52.

17. Flowers bilateral. Adam and Eve *(Aplectrum)*, 99.

L. Leaves not chiefly basal.

18. Flowers bilateral.

O. Lower petal (lip) inflated, slipper-like.
Yellow lady slipper *(Cypripedium)*, 76.

O. Lower petal (lip) flat not slipper-like, fringed.
Yellow fringed orchid *(Blephariglottis)*, 81.

18. Flowers radial.

P. Leaves 3 or more in whorl, at least on lower part of stem.

19. Leaf-blades slender-stalked, plant vine-like.
Wild yam *(Dioscorea)*, 70.

19. Leaf-blades sessile, plant not vine-like.
Yellow trillium *(Trillium)*, 63.

P. Leaves not whorled.

20. Flowers over 1¼ in. long.

Q. Petals and sepals borne on top of ovary.
Blackberry lily *(Gemmingia)*, 71.

Q. Petals and sepals borne beneath ovary.
Yellow or orange lilies *(Lilium)*, 48.

20. Flowers under 1¼ in. long.

R. Leaf-blades with rough margins.
False bell-wort *(Oakesiella)*, 58.

R. Leaf-blades without rough margins.

21. Leaf-blade base surrounding stem.
Twin-seed *(Disporum)*, 56.

21. Leaf-blade base not surrounding stem.
    - S. Perianth parts united. Colic-root *(Aletris)*, 64.
    - S. Perianth parts separate. Bellwort *(Uvularia)*, 57.

14. Petals 4.
- T. Sepals and petals borne above the ovary on slender calyx tube. Evening primroses *(Kneiffia)*, 270.
- T. Sepals and petals borne below the ovary.
  - 22. Stamens numerous, over 10. St. Peter's-wort *(Ascyrum)*, 248.
  - 22. Stamens 2. Horse mint *(Monarda)*, 371.
  - 22. Stamens 4. American columbo *(Frasera)*, 320.
  - 22. Stamens 5. Jewel weed *(Impatiens)*, 233.
  - 22. Stamens 6.
    - U. Flowers bilateral. Colic-weed *(Capnoides)*, 160.
    - U. Flowers radial.
      - 23. Leaves simple.
        - V. Leaves strongly toothed. Water cress *(Roripa)*, 163.
        - V. Leaves scarcely toothed. Worm seed mustard *(Erysimum)*, 164.
      - 23. Leaves pinnately divided to the mid-rib, the terminal lobe larger than the side ones.
        - 23a. Stems hairy. Water cress *(Roripa)*, 163.
        - 23a. Stems smooth or nearly so.
          - 23b. Fruit constricted between the seeds. Charlock *(Sinapsis)*, 162.
          - 23b. Fruit slender not constricted.
            - 23c. Leaflets entire or nearly so. Yellow cress *(Barbarea)*, 161.
            - 23c. Leaflets definitely and finely toothed. Wild or black mustard *(Brassica)*, 162.

14. Petals 5; sepals to be regarded as petals, if petals are not present.
- W. Petals separate.
  - 24. Flowers radial.
    - X. Leaves simple, sometimes deeply lobed.
      - 25. Stems twining.
        - Y. Flowers large with prominent radiate-fringed crown (whorl of parts above the corolla). Yellow passion flower *(Passiflora)*, 256.
        - Y. Flowers small, bearing 3-angled fruit. False buckwheat *(Tiniaria)*, 105.
      - 25. Stems not twining.
        - Z. Stamen stalks united into a slender tube around style.
          - 26. Leaf blades ovate or nearly so. False mallow *(Sida)*, 246.

26. Leaf blades heart-shaped.
Velvet leaf *(Abutilon)*, 245.

Z. Stamen stalks not forming slender tube.

27. Stamens over 10.

a. Leaf-blades deeply 3-7 palmately lobed.
Buttercups *(Ranunculus)*, 145.

a. Leaf-blades entire.

28. Leaves alternate.

b. Calyx apparently absent; perianth of one set of petal-like parts; leaves without glands.
Frost-weed *(Helianthemum)*, 252.

b. Calyx present; flowers with extra set of 5 in-curved parts borne above the petals.
Butterfly weed *(Asclepias)*, 328.

28. Leaves opposite.

c. Leaves with blades, not reduced to scales.
St. John's-wort *(Hypericum)*, 249.

c. Leaves without blades, reduced to minute scales.
Pineweed *(Sarothra)*, 250.

27. Stamens 10 or less.

d. Leaf-blades not fleshy, pistil of 5 united carpels.

29. Flowers in an umbel.
Milkweeds *(Asclepias)*, 328.

29. Flowers not in an umbel but a loose terminal cluster. Wild flax *(Cathartolinum)*, 234.

d. Leaf-blades fleshy, 5 carpels separate.

30. Leaf-blades flat, slender wedge-shaped.
Roseroot *(Rhodiola)*, 177.

30. Leaf-blades cylindric. Stonecrop *(Sedum)*, 179.

X. Leaves compound.

31. Flowers strictly in umbels.

e. Leaves mostly 3-divided.

32. Ribs of fruit obscure. Alexanders *(Zizia)*, 284.

32. Ribs on fruit, prominent, winged.
Meadow parsnip *(Thaspium)*, 292.

e. Leaves with more than 3 divisions.

33. Fruit with hooked bristles.
Snake-root *(Sanicula)*, 277.

33. Fruit not prominently ribbed.
Pimpernel *(Taenidia)*, 287.

31. Flowers never in umbels.

f. Leaves chiefly of 3 leaflets.

34. Creeping or running plants.
  g. Calyx without leafy bracts between the sepals.
   Barren strawberry *(Waldsteinia)*, 196.
  g. Calyx with leafy bracts between the sepals.
   Indian strawberry *(Duchesnea)*, 194.

34. Stems erect.
  h. Stamens 10. Yellow wood sorrel *(Xanthoxalis)*, 237.
  h. Stamens more than 10.
   Cinquefoil *(Potentilla)*, 195.

f. Leaves chiefly of 5 leaflets, palmately arranged.
 Cinquefoil *(Potentilla)*, 195.

f. Leaflets (at least basal ones) more than 5, pinnately arranged.
 35. Stamens 10. Senna *(Cassia)*, 202.
 35. Stamens more than 10.
  i. Leaves with small leaflets between the larger leaflets.
   Agrimony *(Agrimonia)*, 199.
  i. Leaves without small leaflets between larger leaflets.
   36. Styles bayonet shaped, 2-angled.
    Avens (Geum), 197.
   36. Styles not bayonet shaped.
    Mountain avens *(Sieversia)*, 198.

24. Flowers bilateral.
 j. Leaves simple.
  37. Flowers solitary on slender stalks. Violets *(Viola)*, 254.
  37. Flowers in racemes.
   k. Fruit pods not inflated.
    Hairy fruited bean *(Dolicholus)*, 222.
   k. Fruit pods inflated. Rattle box *(Crotalaria)*, 206.
 j. Leaves compound.
  38. Leaflets 3.
   l. Flowers close clustered in axils of upper leaves.
    39. Plants over 1½ ft. Bush clover *(Lespedeza)*, 221
    39. Plants under 1½ ft.
     Pencil flower *(Stylosanthes)*, 219.
   i. Flowers in terminal or lateral elongated racemes, not close clustered.
    40. Pods rounded. Wild indigo *(Baptisia)*, 205.
    40. Pods flat.
     m. Stamens distinct. Bush pea *(Thermopsis)*, 204.
     m. Nine stamens united, 1 free.
      Hairy fruited bean *(Dolicholus)*, 222.

38. Leaflets more than 3, pinnately arranged.
  n. Leaves terminated with paired leaflets.
    41. Plants not over 2 ft., flowers solitary borne above leaf axils. Sensitive pea *(Chamaecrista)*, 203.
    41. Plants over 2 ft., flowers in axillary racemes.
      o. Large petal of flowers over ½ in. in dia. Pea tree *(Sesban)*, 211.
      o. Large petal of flowers under ½ in. in dia. Giant pea *(Glottidium)*, 212.
  n. Leaves terminated with a single leaflet.
    42. Flowers tinged with purple and pink. Goat's rue *(Cracca)*, 210.
    42. Flowers not tinged with purple and pink. Joint weed *(Aeschynomene)*, 217.

W. Petals united.
  43. Flowers radial.
    p. Petals numerous (over 15) united into funnel-shaped tube. Prickly pear *(Opuntia)*, 257.
    p. Petals 5.
      44. Leaves alternate.
        q. Flowers solitary. Ground cherry *(Physalis)*, 344.
        q. Flowers in a spike.
          45. Corolla tubes short, not longer than sepals. Mullen *(Verbascum)*, 383
          45. Corolla tubes twice as long as the sepals. Gromwell *(Lithospermum)*, 351.
      44. Leaves opposite or whorled.
        r. Stamens 4.
          46. Flowers in terminal clusters. American columbo *(Frasera)*, 320.
          46. Flowers solitary in upper leaf axils, scattered. Mullen foxglove *(Afzelia)*, 397.
        r. Stamens 5.
          47. Sepals and petals borne above the ovary. Horse gentian *(Triosteum)*, 424.
          47. Sepals and petals borne below the ovary.
            s. Stamen stalks united at base. Loosestrife *(Lysimachia)*, 310.
            s. Stamen stalks not united at base. Sterile stamen loosestrife *(Steironema)*, 311.
  43. Flowers bilateral.
    t. Anther bearing stamens 2.
      48. All leaves sessile. Hedge hyssop *(Gratiola)*, 391.

48. Leaves stalked.
  u. Stems square. Horse mint *(Monarda)*, 371.
  u. Stems cylindric. Large stone-root *(Collinsonia)*, 381.
t. Anther bearing stamens 4.
  49. Leaves deeply cut into numerous fine segments.
    v. Flowers in terminal spike. Lousewort *(Pedicularis)*, 404.
    v. Flowers solitary, scattered in upper leaf axils. Mullen foxglove *(Afzelia)*, 397.
  49. Leaves simple, not deeply cut.
    w. Sterile filament present in upper side of corolla tube. Figwort *(Scrophularia)*, 385.
    w. No sterile filament present. Giant hyssop *(Agastache)*, 361.
t. Anther bearing stamens 5. False fox glove *(Dasystoma)*, 398.

Flowers Blue

A. Flowers in heads.
  1. Flowers all tubular. Teasel *(Dipsacus)*, 426.
  1. Central flowers tubular, marginal ones strap-shaped. Blue Asters *(Aster)*, 448.
  1. All flowers strap-shaped.
    B. Heads over 1 in. broad. Chicory *(Cichorium)*, 495.
    B. Heads less than 1 in. broad. Blue wild lettuce *(Lactuca)*, 491.
A. Flowers not in heads.
  2. Petals apparently 2, third one reduced. Day-flower *(Commelina)*, 29.
  2. Petals 3; sepals often of same number, shape or color as petals (iris family and others).
    C. Petals and sepals attached below the ovary.
      3. Petals and sepals separate. Wild hyacinth *(Quamasia)*, 51.
      3. Petals and sepals united into a tube. Grape hyacinth *(Muscari)*, 49.
    C. Petals and sepals borne above the ovary.
      4. Flowers large, over 1 in. dia. Iris *(Iris)*, 73.
      4. Flowers small, under 1 in. dia. Blue-eyed grass *(Sisyrinchium)*, 72.
  2. Petals 4.
    D. Petals separate. Larkspur *(Delphinium)*, 136.
    D. Petals united.
      5. Flowers under ¾ in. long.
        E. Sepals and petals borne above the ovary. Bluet *(Houstonia)*, 419.

E. Sepals and petals borne below the ovary.
6. Flowers many in a dense slender conical raceme; plants over 2 ft. Culver's root *(Leptandra)*, 396.
6. Flowers few, scattered in an open spike or raceme; plants under 2 ft. Speedwell *(Veronica)*, 395.
5. Flowers over ¾ in. long.
F. Petals fringed. Fringed gentian *(Anthopogon)*, 319.
F. Petals not fringed. Ague-weed *(Gentianella)*, 317.
2. Petals 5 or more.
G. Petals separate.
7. Flowers radial.
H. Plants trailing; leaf-blades lobed, thin.
Passion-flower *(Passiflora)*, 256.
H. Plants erect; leaf-blades fleshy, cylindric.
Fame-flower *(Talinum)*, 113.
7. Flowers bilateral.
I. Carpels few, separate.
8. Flowers spurred. Larkspur *(Delphinium)*, 136.
8. Flowers not spurred but hooded.
Monkshood *(Aconitum)*, 137.
I. Carpels 1.
9. Flowers under 1 in. long, leaflets over 3.
J. Leaflets palmately arranged. Lupine *(Lupinus)*, 207.
J. Leaflets pinnately arranged; terminal leaflet a tendril.
Vetch *(Vicia)*, 230.
9. Flowers over 1 in. long, leaflets 3.
K. Styles hairy only at tip; flowers not over 1½ in. long.
Virginia butterfly pea *(Bradburya)*, 226.
K. Styles hairy on inner side, flowers about 2 in. long.
Butterfly pea *(Clitoria)*, 227.
G. Petals united.
10. Flowers radial.
L. Leaves opposite.
11. Stamens 2.
M. Flowers in a dense terminal conical spike.
Culver's root *(Leptandra)*, 396.
M. Flowers solitary or in leafy open racemes or spikes.
Speedwell *(Veronica)*, 395.
11. Stamens 4.
N. Upper 3 lobes of calyx much longer than lower 2.
Blue-curls *(Trichostema)*, 359.
N. Lobes of calyx equal or nearly so.
12. Flowers in spikes.

O. Fruiting calyx strongly contracted above.
Blue-hearts *(Buchnera)*, 401.

O. Fruiting calyx not contracted above.
Vervain *(Verbena)*, 354.

12. Flowers 1-several from upper leaf axils.
P. Flowers over ½ in. long.
Twin blue-bells *(Ruellia)*, 407.

P. Flowers under ½ in. long.
False pennyroyal *(Isanthus)*, 358.

11. Stamens 5.
Q. Leaves pinnate.
13. Corolla lobes spirally twisted in the bud.
R. Appendages occurring between calyx lobes.
Creeping water-leaf *(Nemophila)*, 338.

R. Appendages not occurring between calyx lobes.
Large calyx water-leaf *(Macrocalyx)*, 339.

13. Corolla lobes merely overlapping in the bud.
Small-leaved water-leaf *(Phacelia)*, 340.

Q. Leaves simple.
14. Corollas plaited between the lobes.
Ague-weed *(Gentianella)*, 317.

14. Corollas not plaited between the lobes.
Smooth gentian *(Dasystephana)*, 318.

L. Leaves alternate.
15. Leaves pinnate.
S. Styles 1. Greek valerian *(Polemonium)*, 343.
S. Styles 2 or 2-cleft.
16. Corollas lobes spirally twisted in the bud.
Large calyx water-leaf *(Macrocalyx)*, 339.

16. Corolla lobes merely overlapping in the bud.
Small leaved water-leaf *(Phacelia)*, 340.

15. Leaves simple.
T. Sepals and petals borne above the ovary.
17. Corollas bell form. Bell flower *(Campanula)*, 428.
17. Corollas wide flaring, not bell form.
U. Leaf-blades round, sessile.
Venus' looking glass *(Specularia)*, 430.

U. Leaf-blades ovate on petioles.
American bell flower *(Campanulastrum)*, 429.

T. Sepals and petals borne beneath the ovary.
18. Vine-like, twining or trailing.
Wild blue morning-glory *(Pharbitis)*, 334.

18. Not vine-like.
V. Flowers small, in terminal slender racemes.
19. Flowers with bracts at base.
Stickseed *(Lappula)*, 348.
19. Flowers without bracts.
W. Corolla tube twice as long as the calyx.
Heliotrope *(Heliotropium)*, 353.
W. Corolla tube little longer than the calyx.
Forget-me-not *(Myosotis)*, 350.
V. Flowers large, in terminal loose clusters.
20. Corolla tubes slender, bearded within.
Twin-fruit *(Amsonia)*, 324.
20. Corolla tubes broad, bell-like, not bearded within.
Bluebells *(Mertensia)*, 349.
10. Flowers bilateral.
X. Anther bearing stamens 2.
21. Corollas not over ¼ in. long.
Pennyroyal *(Hedeoma)*, 373.
21. Corollas over ¼ in. long. Sage *(Salvia)*, 370.
X. Anther bearing stamens 4.
22. Calyx tubes bearing a curious erect process on upper side.
Skullcap *(Scutellaria)*, 360.
22. Calyx tubes without such appendage.
Y. Plants prostrate or creeping.
23. Leaf-blades kidney shaped or round in outline.
Ground ivy *(Glecoma)*, 364.
23. Leaf-blades ovate.
Trailing dragon-head *(Meehania)*, 362.
Y. Plants erect.
24. Stems branching at base, short branches lying close to ground. Blue toadflax *(Linaria)*, 384.
24. Stems not so branched at base.
Z. Stamens much exserted, one pair much longer than the other. Blue-curls *(Trichostema)*, 359.
Z. Stamens slightly exserted, about equal in length.
False pennyroyal *(Isanthus)*, 358.
X. Anther bearing stamens 5, corolla tube split on upper side.
Lobelia *(Lobelia)*, 431.

Flowers Green or Green Slightly Modified By Other Colors

A. Inflorescence surrounded or enclosed by a modified leaf (spathe).
  1. Leaves basal, very large, simple. Skunk cabbage *(Spathyema)*, 18.
  1. Leaves compound.
    B. Leaflets 3 or 5, club-shape structure of inflorescence included in spathe. Jack-in-the-pulpit *(Arisaema)*, 15.
    B. Leaflets more than 5, club-shape structure exserted. Green dragon *(Muricauda)*, 16.

A. Inflorescence not surrounded or enclosed by modified leaf.
  2. Flowers in heads, with base of heads swollen or broader than upper part. Composite fireweed *(Erechtites)*, 482.
  2. Flowers not in heads.
    C. Petals 3; sepals often of same number, shape or color as petals (lily family and others).
      3. Flowers radial.
        D. Leaves 3 or more in a whorl.
          4. Leaves 3 only, at top of stem under the single flower. Green flowered trillium *(Trillium)*, 63.
          4. Leaves more than 3. Indian cucumber-root *(Medeola)*, 62.
        D. Leaves alternate.
          5. Flowers many in terminal cluster. Hellebore *(Veratrum)*, 40.
          5. Flowers 1 or 2 in the leaf axils scattered along the stem.
            E. Sepals and petals separate.
              6. Sepals and petals ½ in. or under in length. Twisted-stalk *(Streptopus)*, 55.
              6. Sepals and petals over ½ in. long. Twin-seed *(Disporum)*, 56.
            E. Sepals and petals partly united. Solomon's seal *(Salomonia)*, 59.
          5. Flowers solitary at ends of stems. Golden-seal *(Hydrastis)*, 129.
      3. Flowers bilateral.
        F. Leaves basal.
          7. Leaves 1 at base of flower stalk; no leaf present at flowering time.
            G. Lower petal (lip) ¼ in. long. Crane-fly orchis *(Tipularia)*, 97.
            G. Lower petal (lip) but 1/16 in. long. Adder's mouth *(Malaxis)*, 93.

7. Leaves 2 at base of flower stalk.
Large twayblade orchis *(Liparis)*, 94.

F. Leaves not all basal.

8. Flowers short spurred.
Long bracted orchis *(Coeloglossum)*, 79.

8. Flowers not spurred.

H. Flowers 1 or 2 with whorl of reduced leaves below them.
Whorled snake-mouth *(Isotria)*, 87.

H. Flowers more than 2, in a raceme.

9. Stems bearing 2 reduced leaves below the racemes.
Tway-blade *(Ophrys)*, 90.

9. Stems bearing only small scales on upper part.
Southern green orchid *(Ponthieva)*, 91.

G. Petals more than 3; sepals to be regarded as petals if petals are not present.

10. Flowers with ovaries exserted on stalks.
Spurges (*Chamaesyce* and others), 242.

10. Flowers with ovaries not exserted.

I. Leaves basal.

11. Flowers in racemes; carpels 2 or 3, separate, petals below ovary. Alum-root *(Heuchera)*, 184.

11. Flowers in an umbel; carpels 2, united, petals above ovary.
Water pennywort *(Hydrocotyle)*, 275.

I. Leaves not basal.

12. Leaves opposite.

J. Flowers over 1 in. long.
Smooth gentians *(Dasystephana)*, 318.

J. Flowers under 1 in. long.

13. Branches delicate wire-like, regularly much forked.
Forked chickweed *(Anychia)*, 109.

13. Branches stout, not delicate or wire-like.

K. Flowers with extra set of parts (crown) borne above the corolla: inflorescence an umbel.

14. Hoods (units of extra set of parts) containing a horn-like process. Milkweed *(Asclepias)*, 328.

14. Hoods without horn-like process.
Green milkweed *(Acerates)*, 327.

K. Flowers without extra set of parts (crown): inflorescence not an umbel.
Dogbane *(Apocynum)*, 325.

12. Leaves alternate.

L. Stems vine-like, twining or climbing.

15. Stems 4-angled, prickly.
Tear thumb *(Tracaulon)*, 104.

15. Stems not 4-angled, not prickly.
False buckwheat *(Tiniaria)*, 105.

L. Stems not vine-like.

16. Flowers solitary in leaf axils.

M. Flowers bilateral. Green violet *(Cubelium)*, 255.

M. Flowers radial.
Creeping twin-fruit *(Dichondra)*, 331.

16. Flowers in an umbel; fruit armed with hooked spines.
Snake-root *(Sanicula)*, 277.

16. Flowers in racemes.

N. Flowers prominently spurred.
Green larkspur *(Delphinium)*, 136.

N. Flowers not spurred.

17. Membraneous collar-like structure surrounding stems above the leaves.

O. Green corolla-like part of flower 4-lobed.
Virginia knotweed *(Tovara)*, 103.

O. Green corolla-like part of flower 5-lobed.
Smartweed *(Persicaria)*, 102.

17. Without collar-like structure above leaves.
Green milkwort *(Polygala)*, 238.

16. Flowers in a loose or open cluster, not a raceme or an umbel.

P. Carpels more than 1, separate.
Meadow-rue *(Thalictrum)*, 146.

P. Carpels 1 or united into a single pistil.

18. Leaves simple.
Bastard toadflax *(Comandra)*, 417.

18. Leaves compound.
Blue cohosh *(Caulophyllum)*, 155.

II. KEY TO THE TYPICAL SANDHILL WILD FLOWERS

Note: Grouped here are the characteristic flowering herbs of the coarser deep sand soils of the coastal plain. If the plant is not found, try the key dealing with the great forest areas. Perhaps the specimen was collected in a low or flat locality where the water table is temporarily high in which case plant might be found by reference to the key to the savannah wild flowers. The genus description should be consulted before making a decision as to the name.

A. Flowers green. Green milkweed *(Asclepias)*, 328.

A. Flowers white.

1. Flowers in heads.

B. Heads without strap-shaped marginal flowers; all flowers alike.

2. Leaf-blades divided into very fine thread-like segments.
Summer farewell *(Kuhnistera)*, 216.

2. Leaf-blades not finely divided, simple.

C. Heads arranged in a rounded terminal inflorescence; leaf-blade bases not running down stems.
Everlasting *(Anaphalis)*, 456.

C. Heads arranged in a slender, elongated inflorescence; leaf-blade bases extending down stems, wing-like.
Black-root *(Chaenolobus)*, 454.

B. Heads with few ray flowers.
White-topped aster *(Sericocarpus)*, 447.

1. Flowers not in heads.

D. Flowers bilateral; legume type.

3. Plants smooth; corolla yellow fading to white.
Yellow goat's-rue *(Tium)*, 213.

3. Plants more or less hairy; corolla partly colored with red or purple. Goat's-rue *(Cracca)*, 210.

D. Flowers radial.

4. Ovaries exserted on stalks.
Spurges *(Tithymalopsis* and *Chamaesyce)*, 242.

4. Ovaries not exserted on stalks.
  E. Leaves basal. Bear-grass *(Yucca)*, 61.
  E. Leaves not basal.
    5. Leaves opposite.
      F. Leaf-blades narrow, not over ¼ in. wide.
        6. Leaf-blades short, stiff, closely overlapping on stem. Sandhill chickweed *(Alsinopsis)*, 120.
        6. Leaf-blades scattered, not closely overlapping. Spreading wire-plant *(Polypremum)*, 315.
      F. Leaf-blades broader, over ¼ in. wide.
        7. Flowers over ½ in. dia.; leaf-blades not toothed. Sabbatia *(Sabbatia)*, 316.
        7. Flowers under ½ in. dia.; leaf-blades toothed. Sandhill verbena *(Verbena)*, 354.
    5. Leaves alternate.
      G. Plants creeping or vine-like.
        8. Flowers under ¼ in. dia.
          H. Leaf-blades very small, under 1/16 in. long. Sandhill pyxie *(Pyxidanthera)*, 304.
          H. Leaf-blades over ½ in. long. Trailing arbutus *(Epigaea)*, 301.
        8. Flower over 1 in. dia.
          I. Leaf-blades lobed, arrow-like. Wild morning glory *(Ipomoea)*, 334.
          I. Leaf-blades oval or linear. Small-flowered morning glory *(Breweria)*, 332.
      G. Plants not creeping or vine-like.
        9. Plants with stinging hairs; flowers over ¼ in. dia. Tread softly *(Cnidoscolus)*, 241.
        9. Plants without stinging hairs; flowers under ¼ in. dia.
          J. Flowers in racemes. Jointweed *(Polygonella)*, 101.
          J. Flowers in umbels. Scaly seed *(Spermolepis)*, 282.
          J. Flowers in a loose compound, flat-topped cluster. Wire-plant *(Stipulicida)*, 115.

A. Flowers red.
  10. Leaves simple; flowers under ⅛ in. dia. Tragia *(Tragia)*, 239.
  10. Leaves compound; flowers over ¼ in. dia.
    K. Leaflets many, pinnate. Goat's-rue *(Cracca)*, 210.
    K. Leaflets 3. Scarlet pea *(Erythrina)*, 223.

A. Flowers purple.
  11. Plants with milky juice. Milkweed *(Asclepias)*, 328.
  11. Plants without milky juice.

L. Flowers in heads.
  12. Plants with prickly or spiny stems or leaves.
    M. Leaves simple, coarse, very spiny.
      Sandhill thistle *(Carduus)*. 486.
    M. Leaves compound, stem thickly beset with short, recurved spines. Sensitive brier *(Morongia)*. 201.
  12. Plants not prickly or spiny.
    N. Leaves chiefly basal, stem leaves, if present, much reduced.
      13. Leaf-blades broad, lying flat on ground.
        Elephant's foot *(Elephantopus)*, 433.
      13. Leaf-blades narrow to broad, erect or nearly so.
        O. Basal leaves narrow, with rounded ends, few small leaves on stems. False blazing star *(Carphephorus)*, 438.
        O. Basal leaves broader with pointed ends, many small leaves on stems. Vanilla plant *(Trilisa)*, 436.
    N. Leaves not chiefly basal.
      14. Leaves alternate; heads in a spike.
        Blazing star *(Laciniaria)*, 437.
      14. Leaves opposite; heads in a flat-topped cluster.
        Mountain mint *(Koellia)*, 375.
L. Flowers not in heads.
  15. Flowers very small, under ⅛ in. long. Pinweed *(Lechea)*, 253.
  15. Flowers over ⅛ in. long.
    P. Flowers in a compact, slender spike; plants sub-shrubby.
      Lead plant *(Amorpha)*, 215.
    P. Flowers in a raceme; petals 3, sepals 5, stamens 8.
      Milkworts *(Polygala)*, 238.
    P. Flowers otherwise.
      16. Flowers bilateral.
        Q. Leaflets 2-4. Wild sweet pea *(Lathyrus)*, 231.
        Q. Leaflets 3.
          17. Style of flowers bent within twisted keel.
            Wild bean *(Strophostyles)*, 229.
          17. Style not bent within twisted keel.
            Milk pea *(Galactia)*, 224.
      16. Flowers radial.
        R. Petals 3, separate.
          Sandhill spiderwort *(Cuthbertia)*, 27.
        R. Petals 5, united. Southern moss pink *(Phlox)*, 342.
A. Flowers yellow.
  18. Flowers in heads.
    S. Heads with but 1-2 3-petaled flowers blooming at one time.
      Yellow-eyed grass *(Xyris)*, 25.

S. Heads with many flowers.

19. All flowers strap-shaped; leaves chiefly basal.

T. Leaf-blades simple, pubescent, heads few in cluster at top of flowering stalks; outer bracts of heads shorter than inner.
Hawk weed *(Hieracium)*, 489.

T. Leaf-blades smooth, mostly coarsely toothed; outer bracts of heads as long as inner.
Dwarf false dandelion *(Adopogon)*, 494.

19. Marginal flowers strap-shaped, central flowers tubular.

U. Leaves chiefly basal. Leopard's bane *(Arnica)*, 481.

U. Leaves not chiefly basal.

20. Heads mostly under ½ in. dia.

V. Inflorescence flat-topped.
Flat-topped goldenrod *(Euthamia)*, 443.

V. Inflorescence not flat-topped.

21. Heads short stalked, close clustered.
Goldenrods *(Solidago)*, 434.

21. Heads long stalked, very loose or open inflorescence.
Sandhill yellow aster *(Isopappus)*, 442.

20. Heads over ½ in. dia.

W. Chaffy scales between the flowers.

22. Ray flowers sterile, disk flowers fruit producing.
Sunflower *(Helianthus)*, 470.

22. Disk flowers sterile, ray flowers fruit producing.
Woolly-leaf *(Berlandiera)*, 461.

W. Chaffy scales absent between the flowers.
Golden aster *(Chrysopsis)*, 440.

18. Flowers not in heads.

X. Stamens in small flowers on a yellow stemmed spike; carpels 3, united in flowers below the stamen-bearing ones.
Queen's delight *(Stillingia)*, 240.

X. Stamens and carpels not in separate flowers.

23. Flowers bilateral.

Y. Petals separate of legume or pea type of flower.

24. Leaflets 1 (appears as a simple leaf).

Z. Leaflet-blades broadly oval or nearly circular in outline.

25. Stems tufted, spreading on ground.
Rattle box *(Crotalaria)*, 206.

25. Stems 1 or few, erect.
Hairy-fruited bean *(Dolicholus)*, 222

Z. Leaflet-blades narrow, wing-like structure on stems.
Arrow-stemmed rattle box *(Crotalaria)*, 206

24. Leaflets 3.

a. Stamens distinct, ovaries on short stalks and inflated in fruit. Wild indigo *(Baptisia)*, 205.

a. Stamens mostly united, forming tube around the style; ovaries not stalked or inflated in fruit.

26. Flowers purplish at first, becoming yellow in age. Wild bean *(Strophostyles)*, 229.

26. Flowers definitely yellow from the first.

b. Leaflet-blades broadly ovate. Hairy-fruited bean *(Dolicholus)*, 222.

b. Leaflet-blades narrowly ovate or lance-shaped. Pencil-flower *(Stylosanthes)*, 219.

24. Leaflets 2 or 4; flowers with prominent bracts. Hide and seek *(Zornia)*, 218.

24. Leaflets over 9. Goat's-rue *(Cracca)*, 210.

Y. Petals united.

27. Stamens in 2 pairs, equal in length, exserted. Sandhill mullen foxglove *(Afzelia)*, 397.

27. Stamens in 2 pairs, unequal in length, included. Sandhill false foxglove *(Dasystoma)*, 398.

23. Flowers radial.

d. Petals 3; sepals similar in number, color and size.

28. Petals greenish yellow. False aloe *(Manfreda)*, 67.

28. Petals clear yellow. Yellow star grass *(Hypoxis)*, 66.

d. Petals 4.

29. Leaves alternate, sepals and petals borne at end of tube above ovary. Evening primrose *(Kneiffia)*, 270.

29. Leaves opposite, sepals and petals borne below ovary. St. Peter's-wort *(Ascyrum)*, 248.

d. Petals 5.

30. Petals united.

e. Calyx tubular, finally inflated, enclosing fruit. Ground cherry *(Physalis)*, 344.

e. Calyx saucer or cup shaped, not finally enclosing fruit. Nightshade *(Solanum)*, 345.

30. Petals separate.

f. Leaves opposite, blades dotted with minute glands. St. John's-wort *(Hypericum)*, 249.

f. Leaves alternate, blades not dotted.

31. Stamens 5. Wild flax *(Cathartolinum)*, 234.

31. Stamens 12 or more. Frost-weed *(Helianthemum)*, 252.

30. Petals numerous, in 2 series. Prickly pear cactus *(Opuntia)*, 257.

A. Flowers blue.

32. Flowers in heads. Aster *(Aster)*, 448.

32. Flowers not in heads.

g. Flowers radial.

33. Plants trailing or vine-like.
Blue morning glory *(Pharbitis)*, 334.

33. Plants not trailing.

h. Petals 3, with the 3 similar sepals borne above the ovary.

34. Flowers over 1 in. dia. Iris *(Iris)*, 73.

34. Flowers under 1 in. dia.
Blue-eyed grass *(Sisyrinchium)*, 72.

h. Petals united

35. Calyx hairy. Sandhill twin-flower *(Calophanes)*, 406.

35. Calyx smooth.

i. Corolla tubes narrow, petals wide-spreading from end of tube. Sandhill twin-fruit *(Amsonia)*, 324.

i. Corolla tubes funnel-form, wide-flaring.
Sandhill gentian *(Dasystephana)*, 318.

g. Flowers bilateral.

36. Petals separate.

j. Petals apparently 2, 3rd one greatly reduced.
Day-flower *(Commelina)*, 29.

j. Petals 5, flowers of legume or pea type.

37. Flowers solitary.

k. Style slightly hairy at tip; flower not over 1½ in. long.
Virginia butterfly pea *(Bradburya)*, 226.

k. Style hairy on inner side; flower generally over 1½ in. long. Butterfly pea *(Clitoria)*, 227.

37. Flowers in racemes.

l. Flowers over ½ in. long. Lupine *(Lupinus)*, 207.

l. Flowers under ½ in. long.
Sandhill blue-darts *(Psoralea)*, 214.

36. Petals united.

m. Anther bearing stamens 5, exserted from slit in upper side of corolla. Sandhill lobelia *(Lobelia)*, 431.

m. Anther bearing stamens 4.

38. Calyx tube with lobe extending from upper side.
Skull cap *(Scutellaria)*, 360.

38. Calyx tube without lobe on upper side.
Blue curls *(Trichostema)*, 359.

m. Anther bearing stamens 2. Blue sage *(Salvia)*, 370.

III. THE COASTAL DUNE WILD FLOWERS

Note: The coastal dunes over extensive areas are in contact with salt marsh. Certain plants collected in the transition region may be included under the salt marsh key. In case of failure try this latter key.

A. Flowers white.
  1. Flowers in small heads. — Horseweed *(Leptilon)*, 450.
  1. Flowers not in heads.
    B. Flowers over 1 in. long; leaves basal. — Bear grass *(Yucca)*, 61.
    B. Flowers under 1 in. long; leaves not basal.
      2. Petals separate. — Pearlwort *(Sagina)*, 118.
      2. Petals united below.
        C. Petals 4. — Buttonweed *(Diodia)*, 421.
        C. Petals (or sepals) 5 or 6.
          3. Stamens 8. — Seaside smartweed *(Polygonum)*, 102.
          3. Stamens 5. — Seaside heliotrope *(Heliotropium)*, 353.
A. Flowers blue.
  4. Flowers in curved spikes. — Seaside heliotrope *(Heliotropium)*, 353.
  4. Flowers in heads. — Fog-fruit *(Phyla)*, 355.
A. Flowers yellow.
    D. Plant stems spreading close to sand. — Seaside evening-primrose *(Oenothera)*, 270.
    D. Plant stems erect. — Dune ground cherry *(Physalis)*, 344.
A. Flowers purple or red.
  6. Leaves oval or elliptical. — Sea purslane *(Sesuvium)*, 112.
  6. Leaves linear.
    E. Leaves fleshy. — Salt marsh sand spurry *(Tissa)*, 116.
    E. Leaves not fleshy. — Seaside pinweed *(Lechea)*, 253.
A. Flowers greenish, more or less inconspicuous.
    F. Milky juice present. — Dune spurge *(Chamaesyce)*, 242.
    F. No milky juice present. — Tall sea-blite *(Dondia)*, 107.

IV. THE SAVANNAH WILD FLOWERS

Note: If the plant collected cannot be satisfactorily placed with the aid of this key, try the key for the fresh water marsh, a related habitat. If the sandy area in which the plant was collected was of coarse texture, reference to the sandhill key might prove helpful. Final resort should be the main or great forest area key.

A. Flowers white or cream colored.
  1. Flowers in heads.
    B. No flowers strap-shaped; flowers all alike.
      2. Bracts of involucre, long, white, and drooping.
        White bracted sedge *(Dichromena)*, 14.
      2. No such long, white bracts.
        C. Heads single on slender stalks from the ground.
          Hatpins *(Lachnocaulon)*, 26.
        C. Heads numerous in a terminal cluster.
          3. Stems circular as seen in section; involucre present.
            Thoroughwort *(Eupatorium)*, 435.
          3. Stems square as seen in section; involucre absent.
            Savannah mountain mint *(Koellia)*, 375.
    B. Outer or marginal flowers strap-shaped, central flowers tubular.
      Savannah fleabane *(Erigeron)*, 449.
    B. All flowers strap-shaped.
      4. Leaves all basal.
        Night-nodding bog dandelion *(Thyrsanthema)*, 487.
      4. Leaves not basal. White lettuce *(Nabalus)*, 488.
  1. Flowers not in heads.
    D. Petals 3; sepals often of same number, shape or color as petals.
      5. Sepals and petals borne above the ovary.
        E. Flowers radial, solitary or few at top of stem.
          Thread plant *(Burmannia)*, 75.
        E. Flowers bilateral, many in a spike or raceme.
          6. Flowers in a twisted spike.
            Ladies' tresses orchid *(Ibidium)*, 89.
          6. Flowers in an open raceme.

F. Lower petal (lip) fringed.
Bog fringed orchid *(Blephariglottis)*, 81.

F. Lower petal (lip) not fringed.
Savannah white orchid *(Gymnadeniopsis)*, 84.

5. Sepals and petals borne below the ovary.

G. Sepals and petals united forming perianth tube with rough surface. Colic root *(Aletris)*, 64.

G. Sepals and petals not united.

7. Sepals and petals with 1 or 2 glands near base.

H. Sepals and petals with 1 gland; inflorescence branched.
Branched fly poison *(Oceanoros)*, 42.

H. Sepals and petals with 2 glands; inflorescence a simple raceme.

8. Plants smooth. Tall bunch flower *(Zygadenus)*, 43.

8. Plants pubescent.
Tall hairy bunch flower *(Melanthium)*, 44.

7. Sepals and petals without glands.

I. Flowers sessile, the base enclosed in a prominent bract.
Star-flower *(Pleea)*, 35.

I. Flowers stalked, not enclosed in a bract.

9. Leaves numerous (over 15) at base of stem; raceme conic. Turkey-beard *(Xerophyllum)*, 36.

9. Leaves few (under 15) at base of stem; racemes cylindric when mature.

J. Stems pubescent below the inflorescence.
Rough-stemmed bunch flower *(Triantha)*, 34.

J. Stems smooth.

10. Inflorescence over ½ in. in dia.
Savannah fly poison *(Tracyanthus)*, 39.

10. Inflorescence under ½ in. in dia.
False asphodel *(Tofieldia)*, 33.

D. Petals 5.

11. Petals separate.

K. Leaves basal; insect catching plants.

12. Blades over ½ in. wide, with row of prominent bristles along each margin; sensitive.
Venus' fly-trap *(Dionaea)*, 173.

12. Glades under ½ in. wide, covered with glandular appendages. Sundew *(Drosera)*, 172.

K. Leaves not basal; reduced to a stiff pointed structure clasping the stem. Cow-bane *(Oxypolis)*, 294.

11. Petals united.

L. Plants trailing or creeping.

13. Petals united nearly to the top.
  M. Flowers small, under ¼ in. long, stem woody.
  Trailing blueberry *(Vaccinium)*, 303.
  M. Flowers large, over ½ in. long, stem herbaceous.
  Small-flowered morning glory *(Breweria)*, 332.
13. Petals united only near base. Pyxie *(Pyxidanthera)*. 304.

L. Plants erect, not creeping.
14. Stem leaves reduced to small insignificant scales.
Bartonia *(Bartonia)*, 321.
14. Stem leaves not reduced to scales.
  N. Flowers small, under ¼ in. dia.; corolla tubes contracted above. Mitrewort *(Cynoctonum)*, 313.
  N. Flowers larger, over ¼ in. dia.; corolla tubes not contracted above. Sabbatia *(Sabbatia)*, 316.

A. Flowers purple, pink or red.
15. Flowers in heads.
  O. Leaves all basal; flowers strap-shaped, purple only on outside of marginal flowers.
  Night-nodding bog dandelion *(Thyrsanthema)*, 487.
  O. Leaves not basal; flowers all tubular.
    16. Heads in spikes. Bog blazing star *(Laciniaria)*, 437.
    16. Heads in open, rounded, or flat-topped clusters.
      P. Stem leaves numerous, clasping and appressed against stem; leaves pointed. Hound's tongue *(Trilisa)*, 436.
      P. Stem leaves few, leaves chiefly basal, round-tipped.
      False blazing star *(Carphephorus)*, 438.
15. Flowers not in heads.
  Q. Petals 3.
    17. Sepals and petals borne below the ovary.
      R. Flowers solitary, 1 to a plant.
        18. Flowers pink, borne above a whorl of 3 leaves.
        Bog trillium *(Trillium)*, 63.
        18. Flowers purplish blue, borne above a single sword-shaped leaf. Savannah iris *(Iris)*, 73.
      R. Flowers in racemes or spikes.
        19. Plants over 1½ ft. high; flowers white at first, changing to reddish purple in age; perianth parts about equal.
        Bog fly-poison *(Tracyanthus)*, 39.
        19. Plants under 1½ ft. high; flowers not white at first; perianth parts unequal. Milkwort *(Polygala)*, 238.
    17. Sepals and petals borne above the ovary, flowers bilateral.
      S. Flowers solitary on plant or sometimes 2.
      Savannah snake-mouth orchid *(Pogonia)*, 85.

S. Flowers 3 or more on plant.
Grass pink orchid *(Limodorum)*, 96.

Q. Petals 5 or more.

20. Flowers radial.

T. Petals united.

21. Plants creeping close to ground; stems woody; flowers under 1/8 in. wide. Trailing blueberry *(Vaccinium)*, 303.

21. Plants erect; stems herbaceous, flowers over 3/4 in. wide.
Sabbatia *(Sabbatia)*, 316.

T. Petals separate, or nearly so.

22. Leaves all basal, vase-like, insect catching.
Pitcher plants *(Sarracenia)*, 174.

22. Leaves not basal.

U. Flowers with extra whorl of petal-like parts above the petals; in umbels.

23. Flowers red. Red milkweed *(Asclepias)*, 328.

23. Flower greenish purple.
Green milkweed *(Acerates)*, 327.

U. Flowers with no extra whorl of parts above petals; in loose clusters but not in umbels.
Savannah meadow-beauty *(Rhexia)*, 258.

20. Flowers bilateral.

V. Petals united into a tiny slender tube; 2 sepals enlarged, petal-like. Milkworts *(Polygala)*, 238.

V. Petals not forming tiny slender tube; sepals unequal.

24. Flowers solitary in upper leaf axils.
Gerardia *(Gerardia)*, 399.

24. Flowers in spikes.

W. Bracts under flowers shorter than the calyx; corolla tubes slender, petals spreading.
Blue-hearts *(Buchnera)*, 401.

W. Bracts under flowers longer than the calyx; corolla tubes funnel-form. Eared gerardia *(Otophylla)*, 400.

A. Flowers yellow or orange.

25. Flowers in heads.

X. Heads borne on slender leafless stems; leaves basal; flower with 3 petals. Yellow-eyed grass *(Xyris)*, 25.

X. Heads borne on leafy stems.

26. Heads under 3/8 in. dia.

Y. Inflorescence flat-topped.
Rayless flat-topped goldenrod *(Chondrophora)*, 441.

Y. Inflorescence elongated, not flat-topped.
Bog goldenrod *(Solidago)*, 434.

26. Heads over ⅜ in. dia.

Z. Heads without involucre of green bracts surrounding flowers. Red-hot-poker *(Polygala)*, 238.

Z. Heads with involucre of green bracts.

27. Chaffy scales not present among the flowers.

a. Leaves chiefly basal. Leopard's bane *(Arnica)*, 481.

a. Leaves not chiefly basal.

28. Plants smooth or nearly so. Bog sneezeweed *(Helenium)*, 476.

28. Plants conspicuously pubescent. Golden aster *(Chrysopsis)*, 440.

27. Chaffy scales present among flowers.

b. Central flowers of heads arranged in a cone-shaped mass. False cone flower *(Spilanthes)*, 466.

b. Central flowers of heads arranged in a low rounded or flattened mass.

29. Involucral bracts all appressed against the mass of enclosed flowers. Sunflower *(Helianthus)*, 470.

29. Involucral bracts in 2 sets, the outer spreading, the inner appressed. Bog tickseed *(Coreopsis)*, 473.

25. Flowers not in heads.

c. Petals 3, sepals often of same number, shape or color as petals.

30. Inflorescence spreading, flat-topped.

d. Plants smooth, 2 sepals enlarged, petals united into a tube-like structure. Flat-topped milkwort *(Polygala)*, 238.

d. Plants woolly-pubescent, sepals alike, petals not forming tube-like structure, merely united below.

31. Stamens 3. Redroot *(Gyrotheca)*, 74.

31. Stamens 6. False redroot *(Lophiola)*, 65.

30. Inflorescence not flat-topped.

e. Leaves basal.

32. Sepals and petals borne below the ovary; flowers over 1½ in. long. Bog lily *(Lilium)*, 48.

32. Sepals and petals borne above the ovary; flowers under 1½ in. long. Bog yellow star grass *(Hypoxis)*, 66.

e. Leaves not basal.

33. Flowers bilateral; sepals and petals separate, lower petal fringed. Bog fringed orchid *(Blephariglottis)*, 81.

33. Flowers radial; sepals and petals united. Golden colic-root *(Aletris)*, 64.

c. Petals 4.

34. Sepals and petals borne above the ovary; stamens 4; plants not woody below. Marsh evening primrose *(Ludwigia)*, 265.

34. Sepals and petals borne below the ovary; stamens over 10; plants woody below. St. Peter's-wort *(Ascyrum)*, 248.

c. Petals 5.

35. Petals separate.

f. Leaves basal, vase-like with arching cover. Trumpets *(Sarracenia)*, 174.

f. Leaves not basal.

36. Stamens 5.

g. Sepals and petals borne above the ovary. Bog meadow beauty *(Rhexia)*, 258.

g. Sepals and petals borne below the ovary.

37. Ovary 1-celled; leaves alternate. Wild flax *(Cathartolinum)*, 234.

37. Leaves opposite. Loosestrife *(Lysimachia)*, 310.

36. Stamens numerous in 5 clusters. St. John's-wort *(Hypericum)*, 249.

35. Petals united.

h. Leaves basal or apparently absent.

38. Leaves inrolled at edges, covered with sticky mucilage-like substance. Butterworts *(Pinguicula)*, 410.

38. Leaves, if present, minute, not inrolled. Dwarf bladderwort *(Utricularia)*, 411.

h. Leaves not basal.

39. Corolla tubes about 3/8 in. broad, flowers in axils of upper leaves. Mullen foxglove *(Afzelia)*, 397.

39. Corolla tubes narrow, less than 1/4 in. broad; flowers in terminal flat-topped or racemose inflorescence. Milkworts *(Polygala)*, 238.

A. Flowers blue.

40. Flowers in heads.

i. Strap-shaped flowers on margin; central flowers forming flattened or rounded mass. Savannah aster *(Aster)*, 448.

i. No strap-shaped flowers on margin; flowers in cone-shaped mass. Button snake-root *(Eryngium)*, 278.

40. Flowers not in heads.

j. Petals 3; sepals of same number, size and color.

41. Leaves flat, sword shaped, erect.

k. Flowers large, over 1/2 in. dia. Savannah iris *(Iris)*, 73.

k. Flowers under 1/2 in. dia. Blue-eyed grass *(Sisyrinchium)*, 72.

41. Leaves reduced to mere scales on the stems. Thread plant *(Burmannia)*, 75.

j. Petals 5.

42. Corolla tubes split on upper side. Bog lobelia *(Lobelia)*, 431.

42. Corolla tubes not split.

k. Leaves all basal, with incurled edges, sticky.
Butterwort *(Pinguicula)*, 410.

k. Leaves not basal, rough pubescent, opposite.
Blue hearts *(Buchnera)*, 401.

V. KEY TO THE FRESH WATER MARSH WILD FLOWERS

Note: Marshes are always transitional at their borders into communities of other kinds. In case of failure the collector may try one of the other keys which on the basis of the description might prove to be the right one. Especially those covering the savannah and great forest areas might prove helpful. Before deciding upon the plant's name, turn to the genus description.

A. Flowers blue.
 1. Flowers in heads. Aster *(Aster)*, 448.
 1. Flowers not in heads.
  B. Leaves chiefly basal.
   2. Ovaries prominently 3-angled, the angles bearing wings. Thread-plant *(Burmannia)*, 75.
   2. Ovaries' angles not winged. Iris *(Iris)*, 73.
  B. Leaves not chiefly basal.
   3. Flowers in spikes or racemes.
    C. Leaf-blades heart-shaped. Pickerel-weed *(Pontederia)*, 31.
    C. Leaf-blades not heart-shaped.
     4. Stems creeping. Wild for-get-me-not *(Myosotis)*, 350.
     4. Stems erect. Lobelia *(Lobelia)*, 431.
   3. Flowers solitary.
    D. Plants vine-like, climbing. Marsh bluebell *(Viorna)*, 140.
    D. Plants not vine-like, creeping or floating.
     5. Leaves alternate. Blue mud plantain *(Heteranthera)*, 30.
     5. Leaves opposite.
      E. Corollas bilateral, 2-lipped. Blue hedge-hyssop *(Septilia)*, 390.
      E. Corollas radial, or nearly so. Water-hyssop *(Monniera)*, 389.
    D. Plants not vine-like, stems erect or nearly so.
     6. Petals separate, 2 upper large, 1 lower reduced. Day-flower *(Commelina)*, 29.

6. Petals united.
  - F. Flowers radial. Marsh water-leaf *(Nama)*, 341.
  - F. Flowers bilateral. Skull-cap *(Scutellaria)*, 360.

A. Flowers yellow.

7. Flowers in heads.
  - G. Flowers 1-3, projecting from oval overlapping bracts. Yellow-eyed grass *(Xyris)*, 25.
  - G. Flowers many, surrounded by bracts of an involucre.
    - 8. Heads, under ½ in. dia., bracts of involucre in several series. Marsh goldenrod *(Solidago)*, 434.
    - 8. Heads over ½ in. dia., bracts of involucre in 2 series.
      - H. Lower leaves opposite.
        - 9. Seeds (achenes) bearing 2 barbed appendages (awns). Marsh beggar-ticks *(Bidens)*, 474.
        - 9. Seeds with appendages not barbed. Marsh tick-seed *(Coreopsis)*, 473.
      - H. Leaves alternate. Sneezeweed *(Helenium)*, 476.

7. Flowers not in heads.
  - I. Flowers bilateral.
    - 10. Petals united.
      - J. Leaf-blades much divided into slender lobes or leaves sometimes absent. Bladderwort *(Utricularia)*, 411.
      - J. Leaf-blades not divided.
        - 11. Leaves opposite. Hedge-hyssop *(Gratiola)*, 391.
        - 11. Leaves alternate. Yellow mud plantain *(Heteranthera)*, 30.
    - 10. Petals separate.
      - K. Lower petal (lip) fringed. Yellow fringed orchid *(Blephariglottis)*, 81.
      - K. Lower petal (lip) not fringed. Marsh green orchis *(Perularia)*, 82.
  - I. Flowers radial.
    - 12. Inflorescence a slender spike. Golden-club *(Orontium)*, 19.
    - 12. Inflorescence not a spike.
      - L. Petals four.
        - 13. Sepals and petals borne below the ovary.
          - M. Leaves nearly circular in outline. Golden saxifrage *(Chrysosplenium)*, 182.
          - M. Leaves otherwise.
            - 14. Carpels many, separate, leaf-blades linear. Mouse-tail *(Myosurus)*, 142.
            - 14. Carpels 2, united, leaf-blades pinnately divided. Water-cress *(Roripa)*, 163.

13. Sepals and petals borne from top of ovary.
N. Leaves opposite, stems trailing.
Long-stalked false loosestrife *(Ludwigiantha)*, 266.
N. Leaves alternate, stems erect.
15. Stamens 4. False loosestrife *(Ludwigia)*, 265.
15. Stamens 8-12. Primrose willow *(Jussiaea)*, 267.
L. Petals five or more.
16. Carpels 2 or more, separate.
O. Leaves chiefly basal. Marsh marigold *(Caltha)*, 130.
O. Leaves not chiefly basal.
Marsh buttercups *(Ranunculus)*, 145.
16. Carpels united into a single pistil.
P. Stamens numerous, over 5.
17. Flowers very large, over 5 in. dia.; stamens not grouped in 5 clusters.
American lotus *(Nelumbo)*, 149.
17. Flowers smaller, under 2 in. dia.; stamens 9 grouped in 5 clusters.
Marsh St. John's-wort *(Triadenum)*, 251.
P. Anther bearing stamens 5.
18. Sepals and petals borne on top of ovary.
False loosestrife *(Ludwigia)*, 265.
18. Sepals and petals borne below the ovary.
Q. Stamen stalks distinct at base; aborted stamens present.
Sterile stamen loosestrife *(Steironema)*, 311.
Q. Stamen stalks united at base; no aborted stamens present. Loosestrife *(Lysimachia)*, 310.
A. Flowers purple or red.
19. Flowers in heads.
R. Leaves 4-6 at node, whorled; heads solitary, terminal.
Purple-head waterweed *(Sclerolepis)*, 434.
R. Leaves opposite; heads long-stalked from leaf axils.
Water willow *(Dianthera)*, 409.
R. Leaves alternate; heads many in terminal cluster.
Marsh fleabane *(Pluchea)*, 453.
19. Flowers not in heads.
S. Petals 3; sepals often of same number, shape or color as petals.
20. Flowers radial.
T. Leaves not basal, small, very numerous on stems.
Water-moss *(Mayaca)*, 24.

T. Leaves basal.

21. Leaf-blades sword-shaped, over 8 in. long.
Iris *(Iris)*, 73.

21. Leaf-blades narrowly lance-shaped, under 6 in. long, contracted below into long, slender stalks.
Dwarf water plantain *(Helianthium)*, 8.

20. Flowers bilateral.

U. Flowers in a spike; sepals and petals partly united.
Pickerel weed *(Pontederia)*, 31.

U. Flowers in a raceme; sepals and petals separate.
Marsh orchis *(Habenaria)*, 83.

S. Petals 4.

22. Leaves alternate.

22a. Flowers bilateral; leaf-blades 3-nerved from base.
Chaff-seed *(Schwalbea)*, 405.

22a. Flowers radial; leaf-blades not 3-nerved.
Winged marsh loosestrife *(Lythrum)*, 262.

22. Leaves opposite.

V. Stamens 2.

23. Flowers solitary in leaf axils.
Small flowered false pimpernel *(Micranthemum)*, 394.

23. Flowers in clusters, borne on long axillary stalks.
Water-willow *(Dianthera)*, 409.

V. Stamens more than 2.

24. Sepals and petals borne below the ovary; flowers 1 in. or more long.

W. Inflorescence a simple spike-like raceme.
Dragon-head *(Physostegia)*, 367.

W. Inflorescence branched.
Beautiful marsh mint *(Macbridea)*, 366.

24. Sepals and petals borne above the ovary.

X. Wall supporting calyx free from ovary.
Marsh loosestrife *(Ammannia)*, 260.

X. Wall supporting calyx attached to the ovary.
Marsh purslane *(Isnardia)*, 264.

S. Petals 5 or more.

25. Flowers radial.

Y. Petals separate.

26. Stems with collar-like membranes borne above the leaf bases. Marsh smartweeds *(Persicaria, Polygonum)*, 102.

26. Stems without collar-like membranes above the leaf bases.

Z. Leaves opposite; stamens 9.
Marsh St. John's-wort *(Triadenum)*, 251.

Z. Leaves alternate; stamens 5.
Winged marsh loosestrife *(Lythrum)*, 262.

Y. Petals united, at least below.

27. Plants with milky juice.
Marsh milkweed *(Asclepias)*, 328.

27. Plants without milky juice.

a. Corolla throats closed by hairs; leaves opposite.
Mitrewort *(Cynoctonum)*, 313.

a. Corolla throats not closed by hairs; leaves alternate.
Marsh water-leaf *(Nama)*, 341.

25. Flowers bilateral.

b. Flowers brilliant red.

28. Bracts associated with flowers also colored red.
Painted cup *(Castilleja)*, 402.

28. Bracts not red. Cardinal flower *(Lobelia)*, 431.

b. Flowers purplish.

29. Stems square.

c. Calyx 2-lipped. Beautiful marsh mint *(Macbridea)*, 366.

c. Calyx not 2-lipped, 5 lobes about equal.

30. Flowers in dense axillary clusters.

d. Stamens 2. Water hoarhound *(Lycopus)*, 378.

d. Stamens 4. Bittermint *(Mesosphaerum)*, 382.

30. Flowers in elongated spike-like racemes, showy.
Dragon head *(Physostegia)*, 367.

29. Stems not square.

e. Leaves basal, the blades covered with slimy, sticky substance. Butterwort *(Pinguicula)*, 410.

e. Leaves not basal or covered with sticky substance.

31. Anther-bearing stamens 2.
Water-willow *(Dianthera)*, 409.

31. Anther-bearing stamens 4.

f. False or sterile stamens 1, bearded.
Marsh beard-tongue *(Penstemon)*, 387.

f. False stamens 2, each 2-lobed at end.
False pimpernel *(Ilysanthes)*, 393.

f. False stamens none.

32. Leaves alternate. Chaff-seed *(Schwalbea)*, 405.

32. Leaves opposite.

g. Lower lip of corolla overlapping the upper in bud; leaves not toothed.
Marsh Gerardia *(Gerardia)*, 399.

g. Upper lip of corolla overlapping lower in bud; leaves toothed. Monkey-flower *(Mimulus)*, 388.

31. Anther-bearing stamens 6-8.
Marsh milkwort *(Polygala)*, 238.

A. Flowers white or cream colored.

33. Petals and sepals none, flowers in spikes with nodding tips.
Lizard-tail *(Saururus)*, 100.

33. Petals 3; sepals frequently of same number, size, and color.

h. Flowers bilateral. Fringed orchid *(Blephariglottis)*, 81.

h. Flowers radial.

34. Carpels many, separate.

i. Carpels in a single series or whorl.
Water plantain *(Alisma)*, 7.

i. Carpels in several series, forming heads in fruit.

35. Carpels not sharp pointed, beakless.
Dwarf water plantain *(Helianthium)*, 8.

35. Carpels sharp pointed, beaked.

j. Stamens and carpels in different flowers.
Arrow-head *(Sagittaria)*, 10.

j. Stamens and carpels in the same flower.
Bur-head *(Echinodorus)*, 9.

34. Carpels 3, united into one pistil.

k. Petals united.

36. Stamen stalks not connected by white membrane.

l. Stamens 3. Water star-grass *(Heteranthera)*, 30.

l. Stamens 6. Atamasco lily *(Atamosco)*, 68.

36. Stamen stalks connected below by white membrane.
Spider-lily *(Hymenocallis)*, 69.

k. Petals separate. Water moss *(Mayaca)*, 24.

33. Petals 4.

m. Petals separate.

37. Stems 4-angled. Bed-straw *(Galium)*, 423.

37. Stems not 4-angled.

n. Stamens 4. American burnet *(Sanguisorba)*, 200.

n. Stamens 6. Water cress *(Roripa)*, 163.

m. Petals united.

38. Flowers radial. Bartonia *(Bartonia)*, 321.

38. Flowers bilateral.
Small-flowered false pimpernel *(Micranthemum)*, 394.

33. Petals 5 or more.

o. Petals separate.

39. Sepals and petals borne on top of ovary; flower in umbels.

p. Low plants with creeping stems.

40. Leaf-blades linear. Joint-leaf *(Lilaeopsis)*, 290.

40. Leaf-blades broad.

q. Leaf-blades circular in outline, attached to stalk near center of blade. Water pennywort *(Hydrocotyle)* 275.

q. Leaf-blades broadly arrow-shaped, attached to stalk at base. Intelligence plant *(Centella)*, 276.

p. Plants erect in habit.

41. Leaf-blades dissected into very fine thread-like segments. Bishop weed *(Ptilimnium)*, 289.

41. Leaf-blades divided into numerous, broad leaflets, not thread-like.

r. Plants hairy. Cow-parsnip *(Heracleum)*, 295.

r. Plant smooth.

42. Root crowns when cut through middle show cavities separated by thin partitions. Poison hemlock *(Cicuta)*, 285.

42. Root crowns not showing series of cavities.

s. Leaflet-blades of lower leaves deeply cut. Water parsnip *(Sium)*, 288.

s. Leaflet-blades of lower leaves not deeply cut. Cow-bane *(Oxypolis)*, 294.

41. Leaf-blades simple, linear, tubular. Cow-bane *(Oxypolis)*, 294.

39. Sepals and petals borne below the ovary.

t. Carpels 2 or more, separate. Marsh meadow-rue *(Thalictrum)*, 146.

t. Carpels united into 1 pistil.

43. Leaves basal, flowers solitary, terminal. Grass-of-Parnassus *(Parnassia)*, 181.

43. Leaves not basal, flower in racemes. Water pimpernel *(Samolus)*, 309.

o. Petals united.

44. Flowers radial.

u. Anther bearing stamens 2; low, erect plants.

45. Anther sacs separated at top of stamen stalks; weak, watery herbs. Hedge-hyssop *(Gratiola)*, 391.

45. Anther sacs not separated at top of stalks; rigid herbs. Hairy hedge-hyssop *(Sophronanthe)*, 392.

u. Anther bearing stamens 4; low creeping plants. Water-hyssop *(Monniera)*, 389.

44. Flowers bilateral. (If not found below, see key to mints), 401.

v. Anther bearing stamens 2. Water-hoarhound *(Lycopus)*, 378.

v. Anther bearing stamens 4.

46. Stems square. Germander *(Teucrium)*, 357.

46. Stems not square.
Marsh beard-tongue *(Penstemon)*, 387.
v. Anther bearing stamens 5. Mitrewort *(Cynoctonum)*, 313.
A. Flowers greenish.
47. Inflorescence enclosed by a green leaf-like bract (spathe).
w. Leaves compound, leaflets 3.
Dwarf jack-in-the-pulpit *(Arisaema)*, 15.
w. Leaves simple.
48. Leaf-blades arrow-shaped. Green arrow-arum *(Peltandra)*, 17.
48. Leaf-blades oval, very large. Skunk cabbage *(Spathyema)*, 18.
47. Inflorescence not enclosed by a green bract.
x. Flowers in heads; ray flowers absent.
Marsh fleabane *(Pluchea)*, 453.
x. Flowers not in heads.
49. Petals 3, sepals frequently of same number, size or color.
y. Flowers radial, sepals and petals below the ovary.
Narrow-petaled bunch-flower *(Stenanthium)*, 41.
y. Flowers bilateral, sepals and petals above the ovary.
Green orchis *(Perularia)*, 82.
49. Petals 4, ovary globular, enclosed by a calyx tube bearing the 4 "petals" (sepals). Water purslane *(Didiplis)*, 259.
49. Petals 5 or more.
z. Stems 4-angled, thickly beset with small, weaker spines.
Tear thumb *(Tracaulon)*, 104.
z. Stems not 4-angled with no spines.
50. Cylindrical sheath surrounding stem above the leaves.
Marsh smartweed *(Persicaria)*, 102.
50. No sheath surrounding stem.
Ditch stone-crop *(Penthorum)*, 180.

VI. KEY TO THE SALT MARSH WILD FLOWERS

Note: In areas that are transitional from salt into fresh water, plants from both habitats may be found, hence in case of failure it may be well to try the fresh water marsh key if the water is brackish. The specimen should be checked with the genus description.

A. Flowers in heads.
 1. Plants creeping.
  B. Stems woody, angled, leaf-blades linear. Saltwort *(Batis)*, 111.
  B. Stems herbaceous, not angled, leaf-blades broad.
   Fog-fruit *(Phyla)*, 355.
 1. Plants erect.
  C. Heads not over ½ in. dia.
   2. Heads purplish. Marsh fleabane *(Pluchea)*, 453.
   2. Heads yellow. Salt marsh goldenrods *(Solidago)*, 434.
  C. Heads over ½ in. dia.
   3. Rays purplish. Salt marsh aster *(Aster)*, 448.
   3. Rays yellow. Sea ox-eye *(Borrichia)*, 469.
A. Flowers not in heads.
 4. Leaf-blades linear, or absent.
  D. Plants twining and climbing.
   Salt marsh milkweed *(Seutera)*, 330.
  D. Plants not twining.
   5. Leaves absent, stems fleshy, jointed.
    Samphire *(Salicornia)*, 106.
   5. Leaves present.
    E. Leaves basal.
     6. Flowers blue.
      Salt marsh blue-eyed grass *(Sisyrinchium)*, 72.
     6. Flowers white. Salt marsh arrow-head *(Sagittaria)*, 10.
    E. Leaves not basal.
     7. Leaves alternate.
      F. Leaves borne singly at the nodes.
       Tall sea-blite *(Dondia)*, 107.

F. Leaves clustered at the nodes.
Joint-leaf *(Lilaeopsis)*, 290.

7. Leaves opposite.
G. Petals united. Salt marsh gerardia *(Gerardia)*, 399.
G. Petals separate.
9. Plants creeping. Salt marsh sand spurry *(Tissa)*, 116.
9. Plants erect. Salt marsh loosestrife *(Lythrum)*, 262.

4. Leaf-blades definitely broader than linear.
H. Plants creeping.
10. Flowers in terminal racemes.
Seaside smartweed *(Polygonum)*, 102.
10. Flowers solitary in the leaf axils.
I. Petals united. Water-hyssop *(Monniera)*, 389.
I. Petals separate. Sea purslane *(Sesuvium)*, 112.
H. Plants erect.
11. Leaves basal. Sea lavender *(Limonium)*, 307.
11. Leaves not basal. Salt marsh sabbatia *(Sabbatia)*, 316.

VII. KEY TO THE AQUATIC PLANTS

Note: The true aquatics may be separated from the marsh plants by the fact that no leaves are erect above the surface of the water; the leaves are all of the submersed or floating type. The few submersed plants of the salt water sounds are included here along with the larger number of fresh water species. Refer to the genus description for final checking.

A. Stems or leaves all of floating type.
  1. Small plants floating on surface, not earth rooted.
    B. Plants rootless.
      2. Plant bodies under 1/16 in. dia., nearly globular.
        Dwarf duckweed *(Wolffia)*, 20.
      2. Plant bodies between ¼ and ½ in. long, saber-form.
        Southern dwarf duckweed *(Wolffiella)*, 21.
  B. Plants with water roots.
      C. Roots several. Greater duckweed *(Spirodela)*, 23.
      C. Root solitary, one to each colony. Duckweed *(Lemna)*, 22.
  1. Plants earth rooted.
    D. Circular leaf-blades, not split to center.
        American lotus *(Nelumbo)*, 149.
    D. Circular leaf-blades, split to center.
      4. Flowers yellow. Spatterdock *(Nymphaea)*, 150.
      4. Flowers white. Water lily *(Castalia)*, 151.

A. Leaves of two types, floating and submersed.
  5. Surface leaves toothed or divided.
        Mermaid weed *(Proserpinaca)*, 273.
    E. Leaves strictly alternate.
    E. Leaves frequently whorled.
        Parrot's feather *(Myriophyllum)*, 274.
  5. Surface leaves not toothed or divided.
    F. Leaf stalks attached to blades near center (peltate).
      6. Submersed leaves finely divided.
        Carolina water-shield *(Cabomba)*, 147.
      6. Submersed leaves entire. Frog-leaf *(Brasenia)*, 148.

F. Leaf stalks attached at base of blades.

7. Blades heart-shaped, with deep division at base. Floating heart *(Limnanthemum)*, 323.

7. Blades not heart-shaped.

G. Surface leaves with blades tapering toward base, not oval. Water starwort *(Callitriche)*, 243.

G. Surface leaves with blades oval or elliptic. Pondweed *(Potamogeton)*, 3.

A. Leaves all submersed.

8. Leaves all basal.

H. Leaves long, ribbon-like.

9. Leaf bases inflated. Eel-grass *(Zostera)*, 4.

9. Leaf bases not inflated. Tape-grass *(Vallisneria)*, 11.

H. Leaves not ribbon-like, in tufts. Pipewort *(Eriocaulon)*, 26.

8. Leaves not basal.

I. Leaves opposite or whorled.

10. Leaf-blades simple.

J. Leaves short, linear, under ¾ in. long.

11. Leaves opposite. Water-purslane *(Didiplis)*, 259.

11. Leaves in whorls of 3. Waterweed *(Philotria)*, 12.

J. Leaves linear, thread-form, over ¾ in. long, opposite.

12. Blades without evident teeth. Horned pondweed *(Zannichellia)*, 1.

12. Blades with evident teeth. Water nymph *(Naias)*, 5.

10. Leaf-blades divided.

K. Flower stalks greatly inflated. Featherfoil *(Hottonia)*, 308.

K. Flower stalks not inflated. Hornwort *(Ceratophyllum)*, 128.

I. Leaves alternate.

13. Leaf-blades divided into narrow lobes.

L. Plants attached to stones in running water. River-weed *(Podostemon)*, 175.

L. Plants not attached to stones in running water.

14. Leaves bearing minute bladders (animal traps). Bladderwort *(Utricularia)*, 411.

14. Leaves not bearing bladders. Water crowfoot *(Batrachium)*, 144.

13. Leaf-blades simple.

M. Leaf-blades circular in outline. Frog's-bit *(Limnobium)*, 13.

M. Leaf-blades linear.

15. Leaves over ½ in. long.

N. Leaf-blades very fine, hair-like.

Ditch-grass *(Ruppia)*, 2.

N. Leaf-blades coarse, not hair-like.

Pondweed *(Potamogeton)*, 3.

15. Leaves not over ½ in. long.

Water-moss *(Mayaca)*, 24.

SECTION II

DESCRIPTION OF THE GENERA OF HERBACEOUS WILD FLOWERS WITH KEYS TO THE IMPORTANT SPECIES

DIVISION I: PLANTS WITH FLOWER PARTS IN THREES OR SIXES, LEAVES USUALLY PARALLEL VEINED. (MONOCOTS).

Pondweed Family *(Zannichelliaceae)*

1. **Horned pondweed** *(Zannichellia palustris)*. Delicate aquatic herbs with male and female flowers on same plant, the former with 1 stamen, the latter with 2-6 carpels: leaves linear, 1-nerved, 1-3 in. long, opposite or whorled. In ponds and ditches. Spring and summer.

2. **Ditch-grass** *(Ruppia maritima)*. Very delicate aquatic herbs, the naked flowers with two stamens and 4 carpels in each: leaves linear, thread-like 1-4 in. long, alternate. In quiet water, chiefly eastern. Summer.

3. **Pondweed** *(Potamogeton)*. Slender rooted plants with flowers having 4 stamens and 4 carpels: many species, some with upper leaf-blades elliptical and of floating type, in others they are linear and all submersed. In quiet water. Summer.

Eel-grass Family *(Zosteraceae)*

4. **Eel-grass** *(Zostera marina)*. Flaccid or weak plants with staminate and carpellate flowers, separate on flowering stem: leaf-blades ribbon-like, sometimes as much as 6 ft. long. In shallow salt water. Summer.

FIG. 128. ARROW-GRASS SOCIETIES ARE FREQUENTLY ENCOUNTERED ALONG THE SOUND SHORES.

NAIAS FAMILY *(Naiadaceae)*

5. Water-nymph *(Naias)*. Delicate plants with staminate and carpellate flowers on different plants, or in some species both on the same plant; staminate flowers with but a single stamen, carpellate with one pistil: leaf-blades linear, opposite or whorled. Three species in quiet water. Summer.

ARROW-GRASS FAMILY *(Scheuchzeriaceae)*

6. Arrow-grass (*Triglochin striata,* fig. 128). Smooth plants, 8-12 in. high with flowers in racemes or spikes, very small, yellow to green: leaf-blades erect, slender, slightly fleshy in texture. Salt marshes. Summer and fall.

WATER-PLANTAIN FAMILY *(Alismaceae)*

7. Water-plantain *(Alisma subcordatum)*. Perennial, 1-3 ft. high, with flowers in a very loose terminal cluster; carpels many, separate, arranged in a whorl: leaf-blades oval, long stalked, erect. Marshes. Spring to fall.

8. **Dwarf water-plantain** *(Helianthium parvulum)*. Perennial, not over 6 in. tall with flower stalks spreading from common point at tip of main flower stem; sepals and petals 3 each, carpels few, separate: leaf-blades narrow lance-shaped, terminating long slender petioles. Marshes. Spring to fall.

9. **Bur-head** *(Echinodorus)*. Annuals or perennials with flowers in loose terminal clusters; carpels numerous, crowded on the enlarged receptacle: leaf-blades with somewhat heart-shaped bases, stalked; main veins spread from point of attachment of blade to petiole. Two species in marshes. Summer.

10. **Arrow-head** *(Sagittaria)*. Perennials with flowers in whorls of threes arranged along upper portion of main stalk; staminate flowers above carpellate ones when both are on same plant: leaf-blades mostly arrow-shaped, some oval; all basal. Many species. Marshes. Summer to fall.

TAPE-GRASS FAMILY *(Elodeaceae)*

11. **Tape-grass** *(Vallisneria spiralis)*. Perennial with carpellate flowers at ends of long coiled very slender stalks; staminate flowers are cut off from plant and floating on the surface carry out pollination by making accidental contact with the female flowers: leaf-blades ribbon-like, reaching six feet in length. In quiet water. Late summer.

This plant is also called "wild celery" and is an important duck food.

12. **Water-weed** *(Philotria)*. Annual with flowers on very slender stalks enclosed at base by a two-cleft envelope: leaf-blades narrow, pointed, sessile, in whorls of 3. In shallow quiet water. Spring and summer.

FROG'S-BIT FAMILY *(Hydrocharitaceae)*

13. **Frog's-bit** *(Limnobium Spongia)*. Perennial with 3-parted flowers, the petals linear, the sepals oval; stamen stalks

of staminate flowers united in single column with anthers borne in groups at different heights: leaf-blades nearly round in outline, blunt pointed at tip, stalked, basal. Marshes. Summer.

Sedge Family *(Cyperaceae)*

14. White-bracted sedge (*Dichromena,* fig. 59). A sedge with white bracts or modified leaves which simulate a flower and therefore included among the savannah plants. The true flowers are within the minute overlapping scales of the spikelets aggregated in the center of the "flower." In savannah lands. Summer.

Arum Family *(Araceae)*

15. Jack-in-the-pulpit *(Arisaema).* Perennial with inflorescence enclosed in overarching modified leaf, the whole borne on a stalk arising from the leaf base: leaves 1 or 2, compound, erect, the 3 leaflet-blades ovate. Three species in moist soil. Spring.

The "hot" taste of the short solid underground stem of these plants is due to innumerable microscopic crystalline spicules contained in the cells. These prick the tongue like a thousand miniature pin points. Another curious fact about the jack-in-the-pulpit is that by changing the soil habitat the male plants may be transformed to female ones and vice-versa.

KEY TO SPECIES

Club-structure of inflorescence brown or black. In bogs. *A. pusillum.*
Club-structure greenish. In wet woods. *A. triphyllum.*

16. Green dragon *(Muricauda dracontium).* Perennial, with a solitary compound leaf 1-4 ft. high, the inflorescence with its enveloping leaf-like "spathe" appearing as borne from the leaf base; flowers minute on base of inflorescence stalk which extends above the flowers in the form of a tapering appendage, not club-shaped as in Jack-in-the-pulpit: leaflet-blades elliptical, about 3 times as long as wide, stalked

FIG. 129. LIKE GIANT ARROWS THE LEAVES OF THIS ARUM (ARROW-ARUM) PUSH OUT OF THE WET SOIL IN EARLY SPRING. PHOTO BY A. F. ROLLER.

FIG. 130. APPEARING LIKE SERPENTS CREEPING FROM THE GROUND, THE GOLDEN-CLUBS RAISE THEIR MANY HEADS FROM THE MUD OF THIS POND BORDER. PHOTO BY H. L. BLOMQUIST.

and borne on two spreading branches of the main leaf stalk so that the whole compound leaf is much wider than long. Mostly in wet ground, more common in our western area. Early summer.

17. **Green arrow-arum** (*Peltandra,* fig. 129.) Perennials, 8-20 in. high, with the inflorescence partly surrounded by a modified leaf (spathe) which bends backward rather than overarches: leaf-blades arrow-shaped, very much resembling those of the arrow-heads. Two species in wet ground and marshes. Spring and summer.

18. **Skunk cabbage** *(Spathyema foetida).* Coarse perennial, 1-3 ft. high, with the flowers aggregated in an oval head, 1 in. long, which is enclosed in a 3-6 inch long modified leaf: the very large cabbage-like malodorous leaves will alone distinguish this aroid. Marshes and wet soil. Early spring.

19. **Golden-club** (*Orontium aquaticum,* fig. 130). Perennial with slender yellow spikes borne on stalks which are whitened beneath the flowering portion: leaf-blades large, oval, parallel veined, stalked. Wet ground and marshes. Spring.

Duckweed Family *(Lemnaceae)*

20. **Dwarf duckweed** *(Wolffia Columbiana).* The smallest of the flowering plants with bodies no larger than a pin-head (less than 1/16 in.): flower inconspicuous. Floating aquatic. Summer.

21. **Southern dwarf duckweed** *(Wolffiella Floridana).* Larger (1/4 in.) than the preceding with scythe-shaped plant bodies which become more or less tangled into masses. Floating aquatic. Spring to fall.

22. **Duckweed** *(Lemna).* Small (1/8-3/8 in. dia.) rounded or disc-like plant bodies 1-5 nerved, commonly occurring in temporary clusters of 2 or 3 through vegetative propagation. Floating aquatics. Summer.

FIG. 131. THE PIPEWORTS STAND STIFFLY ERECT LIKE WHITE-HEADED HAT-PINS. PHOTO BY H. L. BLOMQUIST.

23. **Greater duckweed** *(Spirodela polyrrhiza)*. Plant body rounded with 7-12 conspicuous nerves. Floating aquatic which rarely is found in flower.

MAYACA FAMILY *(Mayacaceae)*

24. **Water moss** *(Mayaca)*. Branching and spreading plant with numerous fine overlapping leaves suggesting a moss in general aspect; stamens 3, carpels 3 united; flowers solitary from leaf axils: leaf-blades awl-shaped, translucent. Marshes and pools. Spring to fall.

Yellow-eyed Grass Family *(Xyridaceae)*

25. Yellow-eyed grass (*Xyris,* fig. 132). Mostly perennial herbs with flowers emerging from between closely overlapping brown scales which form a compact cone at top of a naked stem: leaves basal, linear borne in two series on opposite sides of the very short leaf-bearing stem. Bogs, marshes and sandhills, chiefly eastern. Summer.

Pipewort Family *(Eriocaulaceae)*

26. Pipeworts or hatpins *(Eriocaulon, Dupatya* and *Lachnocaulon,* fig. 131). Mostly perennial herbs with tight somewhat flattened heads of tiny white flowers, the heads being borne at the end of stiff slender stalks: leaves all basal, linear, forming tufts. About 9 species distributed in grassy bogs and marshes. Summer.

The larger tussocks of these plants suggest a cushion with hatpins sticking out from it at all angles.

Spiderwort Family *(Commelinaceae)*

27. Sandhill spiderwort *(Cuthbertia).* Perennials 4-24 in. high, with light purplish 3-parted flowers borne in umbel-like clusters, the stalk bases surrounded by a leaf-like bract; stamen stalks hairy: leaf-blades linear or nearly so, arising from stems which spread from the top of the single tap-root. Two species in dry sandy soil. Spring and summer.

28. Spiderwort *(Tradescantia).* Perennials with terminal, purple, blue, or white flower clusters partly enclosed by 2 prominent folded leaf-like bracts; petals 3, broad, showy; stamen stalks and sepals commonly hairy: leaf-blades varying from linear *(T. Virginiana)* to ovate *(T. montana)* in our 2 species, sessile, alternate. In moist sandy soil. Spring and summer.

29. Dayflower *(Commelina).* Mostly perennial species of spreading plants with attractive, apparently 2-petaled flowers in few-flowered clusters which are partly included in a leaf-

FIG. 132. THE DRAGON-FLY INVESTIGATES THE PICKEREL WEED FLOWERS (LEFT). YELLOW-EYED GRASS FLOWERS (RIGHT) BLOOM 1 OR 2 AT A TIME FROM THE SCALY HEAD.

like bract; petals 3, one much reduced, sepals 3, unequal: leaf-blades varying widely in our 6 species. In moist or medium-dry soils. Spring and summer.

PICKEREL-WEED FAMILY *(Pontederiaceae)*

30. Mud-plantain *(Heteranthera)*. Creeping or floating perennials with an inflorescence of white or blue *(H. limosa)*, or yellow *(H. dubia)* flowers arising from an ensheathing bract (spathe); lobes of the corolla narrow, united with the 3 similar sepals at the base; stamens 3, attached to the perianth tube: leaf-blades linear *(H. dubia)* and broadly lance-shaped in *H. limosa*. In mud and floating in shallow water, chiefly eastern. Summer.

31. Pickerel-weed (*Pontederia cordata,* fig. 132). Attractive perennial, 1-4 ft. tall, with showy spikes of blue flowers; sepals and petals pubescent, united into common perianth

FIG. 133. NEAR WHITE LAKE THE SPANISH MOSS FESTOONS THESE TURKEY OAKS.

tube which is 2-lipped, the upper one 3-lobed with 2 yellow spots at base of middle lobe, the lower lip 2-lobed; stamens 6, flowers not all open at same time: leaf-blades narrowly heart-shaped, the stalks sheathing the stems at base. In shallow water and wet ground, abundant eastward. Summer and fall.

PINE-APPLE FAMILY *(Bromeliaceae)*

32. Spanish moss (*Dendropogon usneoides,* fig. 133). Perennial aerial plant with flowers 3-parted, petals separate, yellow: stems and leaves very slender covered with scurfy pubescence; plants aggregated into various sized masses hanging from tree limbs. Summer.

This curious plant (not a moss at all) obtains all it needs from the air—water, carbon dioxide and mineral nutrients, the latter being derived from the dust which falls on the plant. The covering of the plant is high in the water absorption function. The seeds are pointed and sticky which enables

FIG. 134. AMONG THE SMALLEST OF OUR BUNCH FLOWERS IS THE ROUGH-STEMMED ONE, SHOWN HERE WITH ITS NEIGHBOR, THE ORANGE GRASS.

them to remain on limbs to which they fall or are blown. As to habitat, the plants may hang on any kind of tree provided it is located in a region of sufficiently high humidity. Such localities are generally found near water bodies.

BUNCH FLOWER FAMILY *(Melanthaceae)*

33. False asphodel *(Tofieldia glabra)*. Perennial, 1-2 ft. high with a slender terminal raceme under 3/4 in. wide, 1-4 in. long: leaves linear, chiefly basal. Grassy bogs or savannahs. Fall.

34. Rough-stemmed bunch flower (*Triantha racemosa*, fig. 134). Perennial, 1-3 ft. high with racemes 1-4 in. long, with flowers mostly in groups of threes: leaves grass-like, chiefly basal. Grassy bogs or savannahs. Summer and fall.

35. Star flower (*Pleea tenuifolia*, fig. 62). Perennial, 1-2½ ft. tall with flowers in a terminal raceme, each of the 3-10 flowers enclosed in sheathing bract; petals narrow, nearly

FIG. 135. THE FLY-POISON ILLUMINATES THE DEEP SHADE LIKE A TORCH.

linear, pointed: leaves chiefly basal, linear. Grassy bogs or savannahs. Fall.

36. **Turkey beard** *(Xerophyllum asphodeloides)*. Perennial 2-5 ft. high, with showy conical racemes of white flowers as much as 6 in. long; sepals and petals nerved: leaves very narrowly linear, massed at base of stem. Mountain summits and bogs. Spring and summer.

37. **Bunch flower blazing star** *(Chamaelirium luteum)*. Perennial, 1½-2½ ft. high, the racemes with flowers crowded; sepals and petals white, 1-nerved; carpellate and staminate flowers on separate plants: leaf-blades basal, broad above tapering into wide petioles below. Open upland forest. Spring.

38. **Fly poison** (*Chrosperma muscaetoxicum*, fig. 135). Perennial, 1½-4 ft. tall, with racemes 1½ to 2½ in. thick; the white flowers borne on slender pedicels turn green in age; sepals nearly round in outline: strap-like leaves thick at stem base. Upland woods, western. Spring and summer.

39. **Bog fly poison** (*Tracyanthus angustifolius*, fig. 55). Perennial, similar to preceding except the carpels have erect

FIG. 136. THE SMALL-FLOWERED HELLEBORE BEARS THE SEED PRODUCING FLOWERS ABOVE THE MALE OR STAMEN-BEARING ONES.

tips instead of spreading ones and the flowers turn red in age. Grassy bogs, eastern. Spring.

40. Hellebore (*Veratrum,* fig. 136). Perennials with the 6 greenish sepals and petals united to base of the ovary; softly pubescent leaves 4-12 in. long, blades broadly oval in outline. Mountain woods. Summer.

KEY TO SPECIES

Leaves mostly near the base, slightly clasping.	*V. parviflorum.*
Leaves covering most of stem, strongly sheathing.	*V. viride.*

41. Narrow petaled bunch flower *(Stenanthium).* Perennials, with the white flowers numerous in a terminal loose cluster (panicle); sepals and petals narrow, taper-pointed, enclosing base of ovary and attached to it; styles 3, spreading: leaves linear. Mountain woods. Summer and fall.

KEY TO SPECIES

Perianth white.	*S. gramineum.*
Perianth greenish.	*S. robustum.*

FIG. 137. THIS BRANCHED RELATIVE OF THE WELL KNOWN FLY-POISON, IS FREQUENTLY MET ON THE HIGH SLOPES OF OUR MOUNTAINS.

42. Branched fly-poison (*Oceanoros leimanthoides,* fig. 137). Perennial, 1-4 ft. tall, with white flowers or cream colored flowers in branching racemes; sepals and petals with an indefinite gland at base: leaves mainly basal, linear, curving. Mountain woods and grassy bogs, western. Summer.

43. Tall bog bunch-flower (*Zygadenus glaberrimus,* fig. 61). Perennial, 2-4½ ft. high, with cream colored flowers each about 1 in. dia.; petals and sepals narrowed into a stalk-like part below; 2 nectar secreting glands prominent: leaves chiefly basal, blades somewhat inrolled. Grassy bogs, eastern. Summer.

44. Tall hairy bunch-flower *(Melanthium)*. Perennials, 2-5 ft. high, with flowers about 1/2-3/4 in. dia., flower stalks pubescent; petals and sepals contracted below into very prominent stalks; 2 glands present on each: leaves mainly at base, linear.

KEY TO SPECIES

Perianth whitish, petal margins wavy. *M. latifolium.*
Perianth yellowish, petal margin not wavy. *M. Virginicum.*

Onion Family *(Alliaceae)*

45. Wild onion *(Allium, Validallium).* Perennials, readily recognized by the characteristic onion odor; umbels of flowers sometimes replaced by bulblets: leaves slender-cylindric. The species with leaves strictly basal, or not attached to the flowering stalk has been separated under *Vallidallium.* 7 species. Uplands. Spring and summer.

46. False garlic *(Nothoscordum bivalve).* Resembles wild onion but has no onion odor; sepals and petals thin; flowers in umbels on slender pedicels; commonly 2 or more bracts at base of umbel. Uplands, especially sandy ones. Spring.

Lily Family *(Liliaceae)*

47. Day lily *(Hemerocallis fulva).* Perennial, 3-6 ft. tall, the flowers with perianth parts strongly recurved, united below, reddish orange: leaves linear, bulbs clustered and coated. Uplands. Summer.

A garden plant from Europe which though it no longer reproduces by seed, survives indefinitely in abandoned grounds.

48. True lilies *(Lilium).* Perennials with flowers funnel-like, the 6 perianth parts distinct; anthers attached to stalk at middle region: leaves commonly numerous, alternate or whorled. Uplands and bogs.

KEY TO SPECIES

Sepals and petals strongly contracted below into a stalk-like structure.
  Leaves alternate, appressed against stem.
    Savannah lily, (*L. Catesbaei,* fig. 143).
  Leaves whorled below, spreading. Wood lily, *(L. Philadelphicum).*
Sepals and petals not contracted below.
  Sepals and petals not recurved, leaves slightly rough on margins.
    Sepals and petals hardly spreading; flowers horizontal.
      Gray's lily *(L. Grayi).*

FIG. 138. THE TURK'S-CAP LILY IN AN ABANDONED MOUNTAIN FIELD LIFTS ITS STRIKING INFLORESCENCE AGAINST THE DISTANT FOREST.

FIG. 139. THE CAROLINA LILY DEFTLY INTERLOCKS ITS SIX PERIANTH FINGERS

Sepals and petals spreading, flowers nodding.
Wild yellow lily *(L. Canadense).*
Sepals and petals recurved, leaves smooth.
Leaves green, broadest below the middle.
Turk's cap lily (*L. superbum,* fig. 138).
Leaves with bloom (bluish), broadest above the middle.
Carolina Lily (*L. Carolinianum,* fig. 139).

49. Grape hyacinth *(Muscari).* Perennial from bulbs with the single stalk bearing a close raceme of small reflexed deep blue flowers; corolla globose *(M. botryoides)* or more elongated *(M. racemosum),* with 6 small teeth on the margin representing the sepals and petals: leaves linear, basal. Frequent in upland fields, occasionally becoming a pest. Spring.

50. Dogtooth violet *(Erythronium).* Perennials with flowers nodding on slender stalks which arise from a pair of thickish basal, broad, mottled leaves. Upland woods. Spring.

KEY TO SPECIES

Perianth white or purplish within. *E. albidum.*
Perianth yellow. *E. Americanum.*

51. Wild hyacinth *(Quamasia hyacinthina).* Perennial 1-2 ft. high with racemes 3-9 in. long; flowers about as long as their stalks: leaves basal, narrowly linear. Uplands, western. Spring.

LILY-OF-THE-VALLEY FAMILY *(Convallariaceae)*

52. Clinton's lily *(Clintonia).* Perennials with flowers in umbels; leaves mostly 3, basal, blades rather thick, flat, broad. Mountain woods. Spring.

KEY TO SPECIES

Flowers nodding, perianth greenish-yellow. *C. borealis.*
Flowers erect, perianth white. *C. umbellulata,* fig. 140.

53. Wild spikenard *(Vagnera racemosa).* Perennial, 1-3 ft. tall, the white or greenish flowers less than 1/4 in. broad in a panicle at end of the somewhat zigzag stem: leaf-blades

FIG. 140. THE BROAD LEAVES OF CLINTON'S LILY MAY BE EXPECTED ON THE FOREST FLOOR OF HIGH MOUNTAIN SLOPES. PHOTO BY H. L. BLOMQUIST.

broadly oval, sessile on stems, each showing several nerves. Upland woods. Spring and summer.

54. Two-leaved Solomon's seal *(Unifolium Canadense).* Perennial, 2-6 in. high; flowers white, about 1/8 in. dia.; berries speckled with red; stem zigzag: leaf-blades oval nearly straight across the base. Moist upland woods chiefly in the mountains. Spring and summer.

55. Twisted-stalk *(Streptopus)*. Perennials, 1½-3 ft. high, bearing one or two flowers in the leaf axils; petals keeled; stamen stalks short, flat: leaves thin, sessile or clasping. Shady woods, chiefly in the mountains. Spring and summer.

KEY TO SPECIES

Flowers purplish	*S. roseus.*
Flowers whitish tinged with green.	*S. amplexifolius.*

56. Twin-seed *(Disporum)*. Perennials, 1-2½ ft. high, with the flowers pendent from leaf axils; sepals and petals narrow; berries red, seeds commonly 2 in each carpel: leaf-blades somewhat asymmetrical, pubescent or finely so. Upland woods, chiefly in the mountains. Spring.

KEY TO SPECIES

Perianth parts yellow, finely spotted with black.	*D. maculatum.*
Perianth parts greenish yellow, not spotted.	*D. lanuginosum.*

57. Bellwort *(Uvularia)*. Perennials, 6-30 in. high, with stems forking; flowers yellowish, terminal, drooping; sepals and petals, with tips tapering to point and nectaries in their bases bounded by 2 ridges: leaf base surrounding stem. Two species, much alike. Upland woods. Spring.

58. False bellwort *(Oakesiella)*. Perennials, similar to the preceding except flowers are borne opposite to the leaves and the leaves are merely sessile, not surrounding the stem at base. The stems are angled. Two species, much alike. Upland woods and bogs. Spring.

59. Solomon's seal *(Salomonia)*. Perennials, 2-8 ft. high, with greenish-white flowers, 1-8, drooping from leaf axils; sepals and petals partly united; stamens attached to corolla tube: leaf-blades rather thick, oval, sessile: annual stems mostly 2-4 ft.; in *S. commutatum* it sometimes reaches 8 ft. in the mountains; underground stem bears prominent seal-like scars of earlier annual stems. Two species, much alike. Upland woods. Spring and summer.

60. Lily-of-the-valley *(Convallaria majalis).* Perennial with the white flowers drooping in a one-sided raceme; perianth parts united below, recurved above; flowers very fragrant: leaves 2-3, 4-12 in. long, broad and tapering at each end. Mountain woods. Spring.

This species is the same as that of our gardens. It is also native in the mountains of Asia.

Yucca Family *(Dracaenaceae)*

61. Beargrass or **Spanish bayonets** *(Yucca).* Coarse perennials with very tall flowering stems bearing large drooping cream-colored flowers; stamen stalks thickened under the anthers; flowers pollinated by a moth of the same color as the perianth: leaves basal, rigid or recurved, commonly with loose fibers hanging along the margins. Four or five species, chiefly in the coastal plain sandy soils. Summer.

Trillium Family *(Trilliaceae)*

62. Indian cucumber-root *(Medeola Virginica).* Perennial with small greenish flowers in umbels borne just above a whorl of leaf-like bracts: leaves 4-10, all in a single whorl at about the middle of the stem: underground stem with flavor of the cucumber. Upland woods. Spring and summer.

63. Trillium *(Trillium).* Perennials with a single flower from top of stem which latter structure also bears at its apex the three oval leaves in a whorl; sepals green, petals colored. Upland woods. Spring.

KEY TO SPECIES

a. Petals contracted below into narrow stalk-like portions which have a different color from broad part. Petals greenish.
Green wake-robin *(T. viride).*

a. Petals not contracted, uniform in color.
1. Petals greenish.
Southern green wake-robin *(T. discolor* and *Underwoodii).*

FIG. 141. THE TOUCH OF COLOR ON THE PETALS OF THE PAINTED TRILLIUM GIVES THIS SPECIES HIGH RANK AMONG ITS RELATIVES. PHOTO BY H. L. BLOMQUIST.

1. Petals not greenish.
   - b. Petals yellow. Yellow wake-robin *(T. luteum)*.
   - b. Petals chiefly white or sometimes tinged with pink.
     - 2. Petals about 2 in. long, fading to rose color in age. Large flowered trillium *(T. grandiflorum)*.
     - 2. Petals under 1½ in. long.
       - c. Petals of uniform color, pedicel erect. Small flowered white trillium *(T. album)*.
       - c. Petals with purplish marking near base, pedicel recurved. Painted trillium (*T. undulatum*, fig. 141).
   - b. Petals deep purplish, purple brown or crimson.
     - 3. Flowers sessile. Purple wake-robin *(T. sessile)*.
     - 3. Flowers stalked.
       - d. Stamen stalks as long as anthers. Southern purple or red wake-robin *(T. Vaseyi)*.
       - d. Stamen stalks shorter than anthers. Ill-scented wake-robin *(T. erectum)*.
   - b. Petals definitely pink or rose color.
     - 4. Flowers sessile. Huger's trillium *(T. Hugeri)*.

FIG. 142. FLORAL SPIRES ARE TO BE SEEN IN THE SPIKES OF COLIC ROOT (LEFT) AND LADIES' TRESSES ORCHID (RIGHT).

4. Flowers stalked.
 e. Pedicel erect.
  5. Sepals about ½ in. long.
   Pineland wake-robin *(T. pusillum)*.
  5. Sepals ¾ in. long or longer.
   Ill-scented wake-robin *(T. erectum)*.
 e. Pedicel recurved.
  6. Stamen stalks ½ as long as the anthers.
   Nodding trillium *(T. cernuum)*.
  6. Stamen stalks longer than the anthers.
   Mountain wake-robin *(T. stylosum)*.

AMARYLLIS FAMILY *(Leucojaceae)*

64. Colic root (*Aletris,* fig. 142). Perennials, 1½-3 ft. high, with flowers in erect racemes, each flower short stalked; perianth tube roughened exteriorly: basal leaves forming rosette. Two species, the golden *(A. lutea)* and white flowered *(A. farinosa)* colic roots. Damp uplands and grassy bogs. Spring and summer.

65. False redroot *(Lophiola Americana)*. Perennial, 1½-2½ ft. high, with yellowish flowers in a flat-topped terminal

FIG. 143. THE ATAMASCO LILY (LEFT) AND SAVANNAH LILY *(L. Catesbaei)* (RIGHT) HAVE THE PERIANTH OF SIX PARTS BORNE ABOVE AND BELOW THE OVARY RESPECTIVELY. PHOTO (LEFT) BY A. F. ROLLER.

cluster; sepals and petals 3, each united below and attached to the basal half of the ovary; stamens 6 (the similar appearing true redroot, *Gyrotheca,* has only 3) : basal leaves linear, much shorter than the stem, nearly smooth, stem leaves very short, pubescent. In savannah lands, eastern. Spring and summer.

66. Yellow star grass *(Hypoxis hirsuta).* Perennials, 2-6 in. high, with yellow flowers few in cluster at end of flowering stem; petals and sepals pubescent on the outside: leaves grass like; a very narrow-leaved species is reported by Small from Carteret County. Upland woods and grassy bogs. Spring, summer, and fall.

67. False aloe *(Manfreda Virginica).* Perennial, 2-6 ft. tall, with dull yellow flowers in a slender erect raceme; corolla tube funnelform, the six lobes nearly 1/2 in. long: basal leaves many, thick, fleshy. Sandy uplands. Spring and summer.

68. Atamasco lily (*Atamosco Atamasco,* fig. 143). Perennial, with the upright stalk bearing the single white flower like an erect easter lily; perianth borne above the ovary,

$2\frac{1}{2}$-3 in. long: leaves linear, thickish, blunt at the end, all basal. Wet ground, chiefly eastern. Spring.

69. Spider lily *(Hymenocallis rotatum).* Perennial with 2-6 white flowers in a terminal umbel, sepal and petal lobes 2-4 in. long spreading from slender corolla tube of equal length; stamen stalks connected below by white membrane: leaves 1-2 ft. long and 1 in. wide. Wet ground in lower coastal plain, eastern. Spring and summer.

YAM FAMILY *(Tamaceae)*

70. Wild yam *(Dioscorea villosa).* Perennial with twining and branching stems with alternate leaves above a set of whorled leaves near base: flowers minute, in racemes; stamens and carpels in separate flowers: leaf-blades heart-shaped on slender petioles with 9-11 prominent curved veins: underground stem thick, rough. Open upland woods. Summer.

IRIS FAMILY *(Ixiaceae)*

71. Black-berry lily *(Gemmingia Chinensis).* Perennial, 3-4 ft. tall, with showy reddish flowers; perianth mottled, persistent on top of ovary after withering: leaves with their upper faces united above the middle. An escaped plant, a native of China, found along roadsides and borders of woods. Summer.

72. Blue-eyed grass *(Sisyrinchium).* Perennials with small bright blue 6-parted flowers in clusters on the end of a flattened or 2-edged stalk; stamens 3: leaves linear, all basal: commonly low herbs seldom over $1\frac{1}{2}$ ft. high. Many species (15-20) in our area, distinguishable by relatively minor differences. Uplands and grassy bogs. Spring and summer.

73. Blue-flags *(Iris).* Perennials with very showy flowers having the sepals and petals borne above the ovary and partially united; the sepals are bent downward while the petals are commonly erect; styles expanded into 3 spreading petal-like structures with a thin lip near the end beneath the

FIG. 144. THE SAVANNAH IRIS AND TRUMPET LEAVES, A STUDY IN BLUE AND YELLOW.

stigma: leaves sword shaped, interlocking; i. e., one appears to be pushing out from the base of the preceding. Upland woods, grassy bogs and marshes.

KEY TO SPECIES (After Small)

1. Stems short, less than 4 in. tall.
   a. Sepals or outer three lobes showing wrinkled or corrugated surface (crest) at middle region. Dwarf crested iris *(I. cristata)*.
   a. Sepals smooth, crest absent. Dwarf iris *(I. verna)*.
1. Stems tall, over 10 in.
   a. Flowers red or reddish brown. *(I. fulva)*.
   a. Flowers chiefly blue or lilac.
      2. Leaves less than ⅜ in. broad.
         b. Flower solitary; petals reduced, 3 sepals showy with yellow markings; grassy bog species. Savannah iris (*I. tripetala*, fig. 144).
         b. Flowers in clusters, sepals marked with white. Blue gladiole *(I. prismatica)*.

2. Leaves more than ⅜ in. wide.
   c. Seeds in one row in each cavity of ovary or pod.
   Carolina iris *(I. Caroliniana)*.
   c. Seeds in 2 rows.
      3. Flowers on stalks (do not confuse the ovary for a stalk).
      Common blue-flag *(I. versicolor)*.
      3. Flowers sessile or nearly so (without stalks).
      (I. hexagona).

Bloodwort Family *(Haemodoraceae)*

74. Redroot *(Gyrotheca tinctoria)*. Coarse perennial, 1-3½ ft. tall, with very hairy dull yellow flowers in a flat-topped cluster; perianth persists on fruit: leaves iris-like: juice of plant, especially of underground stem, bright red. Swamp borders and grassy bogs, eastern. Spring to fall.

Burmannia Family *(Burmanniaceae)*

75. Thread-plant *(Burmannia)*. Delicate low annuals, 2-5 in. high, with small scales instead of leaves on the stems: flowers 1 or a few together at top of stem; sepals and petals partly united: stems purple below the flower in *(B. biflora)*, absent in *(B. capitata)*. Low grounds and grassy bogs, eastern. Summer and fall.

Orchid Family *(Orchidaceae)*

One of the most remarkable of all plant families in variety of bilateral flower form and color, based upon a common specialized structure. In this family the stamen (for there is usually but one) is united to the style, the stigma appearing between the two anther sacs. The ovary is usually twisted so that the upper petal becomes the lower where it is usually modified into the more or less prominent "lip." The seeds are very small and are germinated with difficulty. A special technique has been worked out in recent years by which the tiny embryo plants may be carried through the critical germination period successfully. This has made possible much better results in the multiplications of the rare and amazing tropical species, under greenhouse conditions.

FIG. 145. IN THE MOCCASIN FLOWERS, CINDERELLA'S SLIPPER IS RESTORED IN A THOUSAND WOODS: PINK LADIES' SLIPPER (LEFT), YELLOW-FLOWERED SPECIES (RIGHT).

76. Ladies' Slipper *(Cypripedium)*. Perennials, 1-2½ ft. tall, with bright colored solitary flowers; lower petal (lip) modified into a sac-like structure, open above: leaf-blades oval or elliptic, sessile. Rich upland woods. Spring.

KEY TO SPECIES

Lower lip longer than the other 2 petals.
The queen's slipper *(C. reginae)*.
Lower lip shorter than the other 2 petals.
Lip pale yellow, not over 1 in. long; flowers fragrant.
Small flowered yellow ladies' slipper (*C. parviflorum*, fig. 145).
Lip rich yellow, lip over ¼ in. long; flowers not fragrant.
Showy yellow ladies' slipper *(C. pubescens)*.

77. Moccasin flower (*Fissipes acaulis*, fig. 145). Perennial, 3-8 in. high, with solitary flowers; lip sac-like, very large, over 1½ in. long; rose-colored: leaves 2, basal. Deep woods, central and western. Spring.

78. Showy orchis *(Galeorchis spectabilis)*. Perennial, 4-12 in. high with purplish flowers in a spike; lip blunt-spurred, other petals and the sepals delicately united to form a hood: leaves 2, blades broadly oval. Upland woods, central and western. Spring.

79. Long-bracted orchis *(Coeloglossum bracteatum)*. Perennial, ½-2 ft. high, with greenish small flowers, 3/8 in. or under, in a spike; lip narrow, lobed at end, other perianth parts converging: erect stems with oval leaves, 4-6 in. long. Upland woods and clearings, western. Spring.

80. Large round-leaved orchis *(Lysias orbiculata)*. Perennial, 1-2 ft. high, with greenish white flowers in a spike; petals narrow, the lower one (lip) about 1/2 in. long and prolonged at base into a slender spur: the 2 leaves with blades oval to circular in outline, lying flat against the ground; the stem leaves are reduced to scales. Upland woods, western. Summer.

81. Fringed orchids *(Blephariglottis)*. Perennials, 1-3 ft. high, with flowers in showy racemes. Lip with spur at base, petals usually toothed or fringed; leaves varying in different species. Uplands, bogs and marshes. Summer.

KEY TO SPECIES

Flowers white. White-fringed orchis *(B. Blephariglottis)*.
Flowers yellow.
  Lip less than ⅜ in. long. Crested yellow-fringed orchis *(B. cristata)*.
  Lip ½ in. long or longer.
    Lip simple, not 3 parted.
      Common yellow-fringed orchis (*B. ciliaris*, fig. 146).
    Lip 3-parted. Ragged-fringed orchis *(B. lacera)*.
Flowers lilac or purple.
  Lobes of the lip merely shallowly cut.
    Fringeless purple orchis *(B. peramoena)*.
  Lobes of the lip deeply cut forming fringe.
    Lip ½-¾ in. long.
      Large purple-fringed orchis *(B. grandiflora)*.
    Lip ⅜ in. long. Small purple-fringed orchis *(B. psycodes)*.

82. Green orchis *(Perularia scutellata)*. Perennial, 1-2 ft. high, with greenish small flowers in a spike; sepals and petals oval, the lip especially broad being only 3/16 in. long; spur about 1/2 length of the ovary: leaves lance-form. Marshes, chiefly eastern. Summer.

FIG. 146. THE YELLOW-FRINGED ORCHIS (LEFT) WOULD BE A GOOD "STATE FLOWER" SINCE IT IS ONE OF THE FEW ATTRACTIVE HERBS DISTRIBUTED IN ALL PARTS OF NORTH CAROLINA. THE WIDE FLARING SEPALS OF THE LARGE SNAKE-MOUTH ORCHID (RIGHT) ARE DISTINCTIVE.

FIG. 147. SAVANNAH WHITE ORCHIDS AND SABBATIA, A STUDY IN WHITE AND ROSE.

FIG. 148. DELICACY AND MODESTY CHARACTERIZE THE NODDING POGONIA, AN UNCOMMON ORCHID OF DEEP WOODS.

83. **Marsh orchis** *(Habenaria Nuttallii)*. Perennial, 1-2½ ft. high, with white flowers in a spike; petals 2-lobed, the lobes being unequal; the spurred lip, 3 parted, only 3/16 in. long: leaves narrow. Marshes, eastern. Summer.

84. **Savannah white orchid** (*Gymnadeniopsis*, figs. 60, 147). Perennials, 1-2 ft. high, with white flowers in a spike; petals smaller than the sepals; lip not fringed: leaves much reduced on upper part of stem. Wet woods, bogs, and marsh borders, chiefly eastern. Summer.

KEY TO SPECIES

Flowers orange-yellow. Yellow spurred bog orchid *(G. integra)*.
Flowers white or greenish.
  Lip 3-lobed; spur club-shaped.
    Green spurred wood orchid *(G. clavellata)*.
  Lip entire; spur very slender. Southern white bog orchid *(G. nivea)*.

85. **Snake-mouth orchid** *(Pogonia)*. Perennials, with purplish flowers solitary; perianth parts except lower petals much

alike: leaves 1 or 2 on stem. Bogs and moist soil, chiefly eastern. Summer.

KEY TO SPECIES

Lower petals bearded; seldom over 1 ft. high.
Small snake-mouth (*P. ophioglossoides*, fig. 56).
Lower petal not bearded; generally over 1 ft. high. Sepals wide spreading. Larke snake-mouth (*P. divaricata*, fig. 146).

86. Nodding pogonia (*Triphora trianthophora*, fig. 148). Perennial, small, delicate orchids; few flowered; sepals and petals much alike, except lower petal: leaves few, blades broadly oval, under 1½ in. long. Rare in shady, moist forests, chiefly in the mountains. Summer.

87. Whorled snake-mouth *(Isotria verticillata)*. Perennial, 10-12 in. high, with the flower solitary or occasionally 2 arising from whorl of 5 leaves at summit of stem; the dark purple sepals very slender, much longer than the greenish yellow petals. Upland woods, western. Spring.

88. Bog arethusa *(Arethusa bulbosa)*. Perennial, 5-10 in. high, with solitary reddish purple flowers; upper 2 petals similar to the sepals and united to them below; lower petal arched: true leaf basal, linear, stem leaves reduced to sheathing scales. Mountain bogs. Summer.

89. Ladies' tresses (*Ibidium*, fig. 142). Perennial orchids having numerous small flowers borne in twisted or spiral spikes; upper petal and the adjacent sepals united: leaves chiefly basal, narrow. Six or seven species of uplands and bogs. Spring, summer, and fall.

90. Twayblade *(Ophrys)*. Rather delicate perennial orchids, 1/2-1 ft. high: flowers small (lower petals or lips about 1/4 in. long) in racemes arising from the pair of broadly ovate or kidney shaped leaves borne midway on the stem. Two species in upland moist woods, western. Spring and summer.

91. Southern green orchid *(Ponthieva racemosa)*. Perennial, 1/2-1½ ft. tall, with small (under 1/4 in.) greenish

FIG. 149. RATTLESNAKE PLANTAIN IS NOT A PLANTAIN BUT AN ORCHID. ITS VEINY LEAF-BLADES ARE VERY DISTINCTIVE. PHOTO BY H. L. BLOMQUIST.

flowers in a raceme; petals drooping, lower petal (lip) smaller than other petals: leaves mostly near the base, blades oval. Rather dry upland woods, central and eastern. Fall to spring.

92. Rattlesnake plantain (*Peramium*, fig. 149). Perennials, 5-20 in. high, with small whitish flowers in spikes;

FIG. 150. THE DELICATE NARROW-LEAVED GRASS-PINK IS A TRUE ORCHID ARISTOCRAT.

upper 2 petals united with the sepal: leaves basal, light in color due to network of white veins; the mottled leaves form characteristic rosettes close to the ground. Two species in upland woods. Summer.

93. Adder's mouth *(Malaxis unifolia)*. Perennial, 4-10 in. high with greenish tiny flowers (lower petal 1/8 in. long) in a spike borne on a slender stem 4-12 in. high; lower petal broad, heart-shaped: leaf-blade oval, solitary near base of stem. Upland woods, western. Summer.

94. Large twayblade *(Liparis liliifolia)*. Perennial, 4-10 in. high, with brownish purple flowers in a raceme; upper 2 petals very narrow, lower petal (lip) broad, nearly 1/2 in. long, with the margin shallowly toothed: leaf-blades broad, commonly in pairs at the base of the stem. Upland woods, western. Spring and summer.

95. Coral-root *(Corallorhiza)*. Perennials, 4-20 in. high, with watery, purplish or reddish stems with leaves reduced to scales; flowers small, in racemes; upper 2 petals smaller than the sepals: underground stem short and much branched forming a coral-like structure. Four species in upland woods, chiefly western. Spring and summer.

96. Grass pinks (*Limodorum,* fig. 150). Perennials with few purplish flowers in a terminal raceme; two upper sepals curved, not equal sided, shorter than the lower one; upper petal (lip in this genus) bearded, the broad bearded portion being borne on a stalk-like base, weakened at one point forming a hinge: leaves mostly reduced to scales, a basal slender one being the functional leaf. Grass bogs or savannahs, chiefly eastern. Spring and summer.

KEY TO SPECIES

Flowers nearly white, only tinged with purple.
Pale grass-pink *(L. pallidum).*

Flowers bright purple.
Leaves very narrow, grass-like.
Grass-leaved pink (L. graminifolium).

Leaves broadly linear becoming slender lance-shape.
Large-flowered grass-pink (*L. tuberosum,* fig. 56.)

97. Crane-fly orchis *(Tipularia unifolia).* Small perennial orchids with the solitary leaf disappearing before the flower stalk develops: the green flowers small, long spurred in a terminal raceme borne on the leafless stalk: leaf-blade oval, commonly spotted with dull purple. Dry woods. Summer.

98. Crested coral-root *(Hexalectris spicata).* Perennial, 8-20 in. high, with the flowers purple-striped; 2 upper petals curved, shorter than the upper sepal; lower petal (lip) with several ridges (crest) running lengthwise: leaves reduced to scales: underground stem coral-like, jointed. Deep mountain woods. Summer.

99. Adam and Eve *(Aplectrum hyemale).* Perennial orchid 1-2 ft. high, with the single large oval leaf disappearing in the fall, so no leaf is present when the flowers appear in spring: flowers dull yellow with purple markings; lower petal nearly 1/2 in. long, 3-lobed: short underground stem commonly has another one attached to it, hence the common name. Upland woods, chiefly in the mountains. Spring.

DIVISION II. PLANTS WITH FLOWER PARTS CHIEFLY IN FOURS OR FIVES OR MULTIPLES OF THESE; LEAVES NETTED-VEINED (DICOTS).

Lizard-tail Family *(Saururaceae)*

100. Lizard-tail (*Saururus cernuus,* figs. 19, 151). Perennial, 2-5 ft. high with the flowers remarkable for the absence of both sepals and petals; the white racemes bending and nodding at the tip: leaf-blades heart-shaped. Wet ground and marshes, chiefly eastern. Summer.

Buckwheat Family *(Polygonaceae)*

101. Jointweed (*Polygonella polygama,* fig. 152). Perennial, 1/2-1½ ft. tall, with small, numerous white or yellow flowers appearing after most of the leaves have fallen; sepals petal-like, corolla absent: leaves narrow tapering toward base, 1/2-1/4 in. long: stem bushy-branched, the branches prominently jointed. Coarse sandy soil, eastern. Summer.

102. Smartweed *(Polygonum* and *Persicaria).* Annuals or perennials with small white, pink or purplish flowers scattered in the leaf axils *(Polygonum)* or borne in terminal racemes *(Persicaria);* perianth of but one set of parts, regarded as a corolla-like calyx; stamens mostly 5-6; ovary 3-angled, more plainly seen in the fruit: leaves various, mostly short and linear, jointed at the base in *Polygonum* and lance-shaped or broader without a jointed base in *Persicaria;* collar-like structure around stem above leaf-base in both genera. Mostly native species, and a few weeds are included here found in a wide range of habitats. Spring, summer, and fall.

103. Virginia knotweed *(Tovara Virginiana).* Flowers similar to the preceding genus except the perianth is always

FIG. 151. EVERY BOTANIST WOULD LIKE TO KNOW WHY THE LIZARD-TAIL SPIKE HAS A WEAK NODDING TIP.

FIG. 152. THE JOINT WEED CAN ESTABLISH ITSELF IN THE DRIEST, COARSEST, WHITE SAND. TO CONSERVE WATER, IT SHEDS ITS LEAVES AT FLOWERING TIME.

4-parted, with 4 stamens within and the small sessile flowers are distantly spaced on a long slender terminal inflorescence axis: stem collars cylindric with fringed margin; leaf-blades broadly ovate with tapering tip, leaves alternate. In upland woods. Summer and fall.

104. Tear thumb *(Tracaulon).* Flowers and inflorescence similar to the preceding genus but varying in color—white, green or red: bases of leaf-blades lobed, arrow-like: stem strongly 4-angled and very prickly with recurved short spines. Moist ground and marshes. Summer.

105. False buckwheat *(Tiniaria).* Annuals and perennials with greenish white flowers and inflorescence similar to smart weed (see above): stem climbing, not prickly: leaf-blades heart-shaped. Three species, much alike in upland woods and thickets. Summer.

Goosefoot Family *(Chenopodiaceae)*

106. Samphire *(Salicornia).* Annuals or perennials, 6-24 in. high, with curious jointed stems bearing minute inconspicuous flowers mostly hidden in the axils of the upper small scale-like leaves: stems fleshy, much branched, the branches erect, green at first turning red: leaves reduced to insignificant scales. Occurring in extensive societies in salt marshes. Summer.

107. Tall sea-blite *(Dondia linearis).* Annual with somewhat fleshy green or purplish stems, 1-3 ft. high, bearing minute flowers clustered in the upper leaf axils; perianth 5 parted, petals absent: leaves linear, tapering at the end with a sharp point. Frequent in salt marshes. Summer.

Whitlow-wort Family *(Corrigiolaceae)*

108. Whitlow-wort *(Paronychia).* Spreading, tufted perennials, the white flowers with sepals petal-like, tipped with bristle-like point; corolla absent: leaves linear 1/2-1½ in. long. On exposed rocks in the mountains.

POKEWEED TO CARPET-WEED FAMILY

KEY TO SPECIES

Flowers closely aggregated in head-like inflorescence.
Mountain whitlow-wort *(P. argyrocoma).*

Flowers in open inflorescence borne on slender wire-like stems.
Nailwort *(P. dichotoma).*

109. **Forked chickweed** *(Anychia).* Low annuals with very small (under 1/8 in. wide) greenish flowers borne in the upper leaf axils; sepals 5, petal-like, petals absent: leaf-blades oval or narrower, sessile, opposite. Two species in open or dry upland woods. Summer.

Pokeweed Family *(Petiveriaceae)*

110. **Pokeweed** *(Phytolacca decandra).* A perennial simulating an annual, 3-10 ft. high, bearing lateral racemes of white or greenish-white flowers: the petal-like structures (sepals since the petals are absent) orbicular, typically 5 in number; stamens typically 10: leaf-blades narrowly ovate or elliptic, stalked. In good soil, on uplands. Chiefly middle and western. Summer.

The young shoots are edible, when boiled. The root is definitely known to be poisonous and the berries are under suspicion.

Salt-wort Family *(Batidaceae)*

111. **Salt-wort** *(Batis maritima).* A low spreading shrub of herb-like aspect bearing the staminate pistillate flowers separately in small axillary cones; the pistillate cones are short-stalked: leaves opposite, the blades narrow, fleshy, flat on one side, curved, not over 1 in. long: branches prominently angled. In and near salt marshes. Summer.

Carpet-weed Family *(Tetragoniaceae)*

112. **Sea purslane** (*Sesuvium,* fig. 153). An annual and a perennial species with flowers solitary in the leaf axils; sepals petal-like, purplish within, each with an appendage on the back; corolla absent: leaf-blades fleshy, broad toward the tip:

FIG. 153. RUGS ON THE TIDAL FLATS ARE FORMED BY THE SEA PURSLANE.

stems spreading flat on the sand. Moist sand, along the coast. Late summer.

PURSLANE FAMILY *(Portulacaceae)*

113. Fame flower *(Talinum teretifolium)*. Perennial, 6-10 in. high, with blue or purple flowers in an open flat-topped cluster borne on slender wire-like stems well above the leafy portion below, sepals 2: leaves narrow-cylindric 1-2½ in. long, fleshy. In thin soil on rocks. Summer.

114. Spring beauty *(Claytonia)*. Perennials with a few white purple-striped flowers in a loose cluster; stamens 5, sepals 2: leaves few, somewhat fleshy, from solid rounded underground stems. Upland woods. Spring.

KEY TO SPECIES

Leaf-blades, narrow, almost linear.
Common spring beauty *(C. Virginica)*.

Leaf-blades broader, approaching oval.
Carolina spring beauty *(C. Caroliniana)*.

Chickweed Family *(Alsinaceae)*

115. Wire plant (*Stipulicida setacea,* figs. 154, 155). Annual, with minute white flowers about 1/16 in. long, scattered over the forking stems: true leaves all basal, disappearing with hot weather: low delicate plants with very slender wire-like stems bearing tiny scales. On coarse white sand, eastern. Late spring.

116. Salt marsh sand spurry *(Tissa marina)*. Annual or biennial with pinkish flowers scattered on upper stems, ovary 1-celled: leaves fleshy, slender-cylindric, 1/2-1¼ in. long on spreading stems; stipules present at leaf base. Salt marshes. Spring.

117. Spurry *(Spergula)*. Annual with slender branches, 6-18 in. high, bearing a loose spreading cluster of small, white flowers; styles 5; sepals longer than the petals: leaves numerous, linear, 1-2 in. long, appearing as whorled at the nodes. An adventive from Europe in old fields and occasionally yards. Summer.

118. Pearl-wort *(Sagina)*. Low spreading massed or matted annuals with delicate stems bearing small greenish white flowers on slender axillary stalks; sepals, petals, and stamens 4 (*S. procumbens* of the dunes) and 5 (*S. decumbens* of inland sandy areas): leaves very slender, thread-form, opposite. In sandy soils along the coast and inland. Spring and summer.

119. Jagged chickweed *(Holosteum umbellatum)*. Low annual with small white slender-stalked flowers in umbel-like clusters; petals 5, shallowly notched; mature open fruiting capsule ending in 6 recurved teeth: leaf-blades ovate, opposite, sessile. In yards and waste areas. Adventive from Europe. Spring.

120. Wild chickweed (*Alsinopsis,* fig. 156). Annuals or

FIG. 154. THIS DELICATE WIRE PLANT SPENDS THE HOT MONTHS IN THE FORM OF SEEDS.

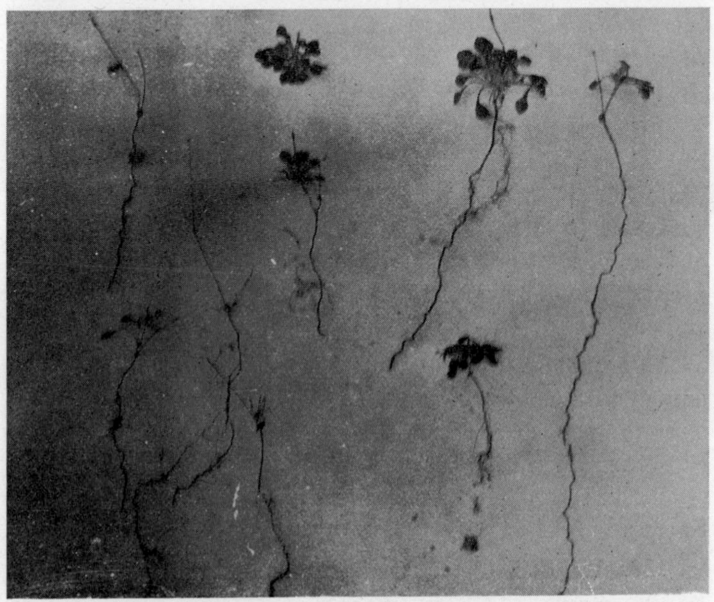

FIG. 155. SPRING ROSETTES OF THE WIRE PLANT WITH THEIR DELICATE ROOTS. THESE DISAPPEAR BEFORE THE PLANT FLOWERS.

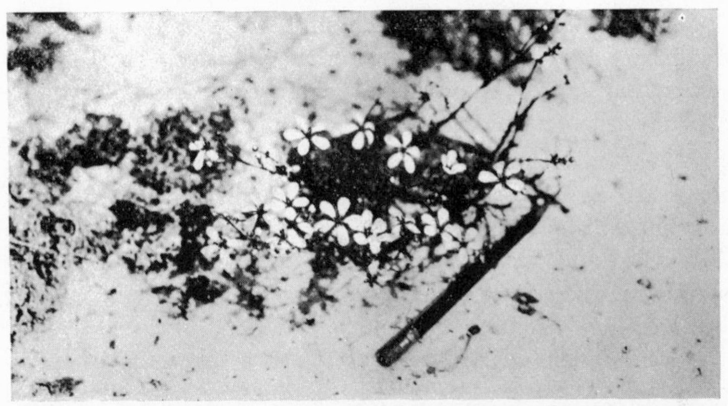

FIG. 156. THE SANDHILL WILD CHICKWEED FLOWERS ARE ALMOST LOST AMID THE WHITE SAND.

FIG. 157. WITH A STARRY RADIANCE, THE GIANT CHICKWEED FLOWERS PEEP OUT IN SHADY PLACES. PHOTO BY H. L. BLOMQUIST.

perennials, 3-9 in. high, with white flowers, petals not notched or slightly so at end; fruit pod 3-valved; ovary 1-celled: leaves opposite, narrow. Upland habitats. Spring and summer.

Six species of which the sandhill *A. Caroliniana* is very distinctive with its closely overlapping, stiff sharp short leaves, resembling moss pink. The flowers are set apart from the leafy portion on slender stalks.

121. Giant chickweed (*Alsine pubera*, fig. 157). Perennial, 4-12 in. high, with showy white flowers; petals 5, deeply notched or cleft, suggesting 10 petals; stamens 10; flowers 3/4 in. across: leaf-blades varying from oval, 1/2-2 in. long: stems spreading from root crown. Shady moist woods, central and western. Spring.

Other species of *Alsine* are the familiar annual chickweeds of which *A. media* is the very common chickweed of the gardens.

122. Mouse-ear chickweed *(Cerastium)*. Low annuals or perennials with small white flowers borne in a loose terminal cluster; petals 5, deeply notched; stamens 10; mature open fruiting capsule terminating in 10 teeth: leaf-blades ovate or nearly so, pubescent in most of our five species, opposite. Waste places. Adventive from Europe. Spring and summer.

Pink Family *(Caryophyllaceae)*

123. Pink *(Dianthus Armeria)*. Annual, 6-18 in. tall with pink and white-dotted flowers in a close terminal cluster surrounded by a whorl of narrow bracts; the notched petal blades are long stalked; calyx tube about 1/2 in. long: leaf-blades linear, the basal ones broader, opposite. In old fields and waste places. Introduced from Europe. Summer.

124. Cow-herb *(Vaccaria Vaccaria)*. Annual, 1-3 ft. tall with light red flowers in a very loose terminal cluster; petal blades stalked; calyx tube 5-angled, the lobes short: leaf-blades varying around lance-shaped, opposite, sessile. In waste areas, a weed from Europe. Summer.

125. Bouncing Bet *(Saponaria officinalis)*. Perennial, 1-3 ft. tall, erect or bending at base, with white or pink flowers in a loose terminal cluster; petal blades appendaged at base, and borne on slender stalks; calyx tube cylindric, 3/4-1 in. long: leaf-blades lance-shaped or elliptical, rather thick, opposite, mostly sessile, 3-veined. In moist upland soils. An introduced plant from Europe, frequently escaping from gardens.

126. Wild pinks *(Silene).* Perennials with showy flowers in terminal clusters; sepals united into a tube with 10 or more nerves; petals generally with an appendage at base of the spreading cleft or divided part: leaf-blades elongate-ovate (varying around lanceolate). Uplands, central and western. Summer.

KEY TO SPECIES

Flowers brilliant crimson.
  Upper stem leaves lance-shape. Fire pink (*S. Virginica,* fig. 158.)
  Upper stem leaves oval.
    Round-leaved fire pink *(S. rotundifolia).*
Flowers white or pink.
  Blades of the petals fringed; white.
    Calyx inflated; leaves in whorls of fours.
      Starry campion *(S. stellata).*
    Calyx not inflated; leaves opposite.
      Mountain campion *(S. ovata).*
  Blades of the petals not fringed; merely shallowly toothed; pink.
    Carolina campion *(S. Caroliniana).*

127. Corn cockle *(Agrostemma Githago).* An attractive annual or biennial weed, 1-3 ft. high, with reddish purple flowers borne singly in the leaf axils; calyx tube 10-ribbed with 5 linear sepals extending beyond the petals; petals 5, the blades shallowly notched: leaf-blades narrow or linear, pubescent, opposite. In fields chiefly in the mountain and piedmont region. An adventive from Europe. Summer.

HORNWORT FAMILY *(Ceratophyllaceae)*

128. Hornwort *(Ceratophyllum).* Flowers not evident: leaves whorled, regularly forking with pointed tips on the ultimate branches. Two species growing submersed in quiet fresh water. Summer.

FIG. 158. THE BRILLIANT RED OF THE FIRE-PINK ALWAYS ATTRACTS ATTENTION.

BUTTERCUP FAMILY *(Ranunculaceae)*

129. Golden seal *(Hydrastis Canadensis)*. Perennial, 8-14 in. high, with greenish-white flowers solitary, terminating the branches; 3 sepals petal-like, corolla none; stalks of stamens, white: leaf-blades lobed and toothed, broader than long. Mountain woods. Spring.

130. Marsh marigold *(Caltha palustris)*. Perennial, 1-2 ft. high, with showy yellow flowers; sepals colored, petals wanting; stamens numerous: leaf-blades kidney-shaped, shallowly toothed: stems rather thick and hollow. Marshes and wet meadows, central and western. Spring.

131. Gold-thread *(Coptis trifolia)*. Perennial, with white flowers solitary on slender stalks from the ground; the sepals petal-like, the petals smaller, club-shaped; stamens many: leaf-blades divided into 3-toothed leaflets attached at end of petiole; veins of leaflets prominent. Mountain woods. Spring and summer.

FIG. 159. THE SKY-SCRAPER OF THE WILD FLOWER CITY; THE TALL FLOWER STEMS OF THE BLACK COHOSH. AT RIGHT IS SHOWN THE COMPLETELY WHITE INFLORESCENCE.

FIG. 160. THE BIZARRE PETALS OF THE COLUMBINE (LEFT) AND THE DUTCHMAN'S BREECHES (RIGHT) GIVE US SOMETHING UNIQUE IN THE MOUNTAIN WOODS. PHOTOS BY A. F. ROLLER.

132. False rue anemone *(Isopyrum biternatum)*. Perennial, with a few white flowers in a loose cluster; sepals 5, petals none; stamens numerous: leaves stalked, branching into threes, with three leaflets terminating the branch ends; leaflets notched and lobed. Moist woods, central and western. Spring.

133. Black cohosh (*Cimicifuga,* fig. 159). Tall perennials, 3-8 ft. high with long white slender racemes; sepals petal-like, petals small, 2-cleft: leaf-blades much divided, the leaflets strongly toothed: rootstalk thick, poisonous. Three species. Upland rich woods, chiefly western. Summer.

134. Baneberry *(Actaea)*. Perennials, 1-2 ft. high with small white flowers in terminal racemes; stamens longer than the petals; carpel 1: leaves compound with toothed leaflets. Rich woods, western. Spring.

KEY TO SPECIES

Fruit red with slender stalks.	Red baneberry *(A. rubra)*.
Fruit white with thick stalks.	White baneberry *(A. alba)*.

135. Columbine (*Aquilegia Canadensis,* fig. 160). Perennial, 1-2 ft. high with showy red flowers, nodding; petals prominently spurred, the spurs with slightly enlarged tips; stamens numerous: leaf-blades three-divided, the leaflets 3-lobed. Rocky woods, chiefly central and western. Spring and summer.

136. Larkspur *(Delphinium)*. Perennials, 1-4 ft. high with blue or white flowers in erect racemes, showy; sepals 5, one bearing the prominent spur; petals 2-4: leaf blades palmately deeply lobed. Six species in mountain woods. Spring and summer.

137. Monkshood (*Aconitum,* fig. 161). Perennials, weak, semi-climbing herbs with attractive, hooded blue or yellowish-white flowers, not spurred; carpels 2-5, stamens many: leaves scattered along the stem, blades palmately deeply lobed. Two species in cool, moist woods, central and western. Summer.

FIG. 161. THE MONKSHOOD OR WOLFS-BANE POSSESSES HIGHLY DECORATIVE LEAVES.

138. Hepatica (*Hepatica,* fig. 162). Low perennials with purple or white flowers having the sepals petal-like and showy, under which are 3 modified and reduced leaves appearing as a calyx; petals none: leaves borne from the underground stem, blades 3-lobed. Two species in upland moist woods. Spring.

139. Anemone *(Anemone).* Perennials with solitary white flowers having 5-18 petal-like sepals; petals absent; stamens many, shorter than the sepals; seeds (mature small carpels or achenes) flattened: leaf-blades palmately lobed and divided into 3 or more leaflets. Woods, chiefly western. Spring and summer.

KEY TO SPECIES

Seeds definitely hairy.
  Stem from ground with but 1 flower.
    Carolina anemone *(A. Caroliniana).*
  Stem from ground branched with more than 1 flower.
    Flower greenish. Green thimbleweed *(A. Virginiana).*

FIG. 162. THE UNFOLDING OF THE DELICATELY TINTED PETALS OF THE LIVERWORT OR HEPATICA IS A SURE SIGN OF SPRING EVERYWHERE. PHOTO BY H. L. BLOMQUIST.

Flower pure white. White thimbleweed *(A. riparia)*.

Seeds smooth or nearly so.

Leaves mostly white, 5 leaflets or divisions.

Wind flower *(A. quinquefolia)*.

Leaves mostly with 3 leaflets. Mountain anemone *(A. trifolia)*.

140. Marsh blue-bell *(Viorna crispa)*. Perennial, delicate, climbing vine with attractive 4 or 5 parted blue or purplish flowers borne solitary in the leaf axils; petals none, the sepals being colored and having wavy margins; carpels many, separate: leaf-blades pinnate, the leaflets mostly in three's, entire, smooth, opposite. In fresh water marshes of the eastern region. Spring and summer.

141. Rue anemone *(Syndesmon thalictroides)*. Perennial, 4 to 10 in. high with few white flowers on slender pedicels from a whorl of leaf-like bracts; sepals petal-like, petals none: leaf-blades 3-divided, the leaflets with slender petioles and coarsely toothed. Upland woods, central and western. Spring.

142. Mouse-tail *(Myosurus minimus).* Annual 2-6 in. high, with greenish yellow flowers having spurred sepals and central cone of many carpels which elongates with age into the "mouse-tail": leaves linear, basal in a tuft. Wet ground, eastern. Spring and summer.

143. False bugbane *(Trautvetteria Carolinensis).* Perennial 2-3 ft. high, with small white flowers in a loose cluster; 3-5 sepals petal-like, petals absent: leaves from the ground blades 5-8 in. broad, but not as long, deeply 5-9 lobed, the margin toothed. Mountain woods. Summer.

144. Water crowfoot *(Batrachium trichophyllum).* Perennial aquatic, with white flowers; sepals and petals 5; flowers 1/2-3/4 in. broad borne singly from the axils of the thread-form, much divided leaves. Ponds and streams. Summer.

145. Buttercup *(Ranunculus).* Perennials with yellow flowers having 5 sepals which fall early and 5 yellow petals, commonly shiny on inner side; each petal has at its base a nectar bearing cavity and a minute scale; carpels many in a cone: leaf-blades palmately lobed and divided. Marsh, moist ground or uplands. Spring and summer.

About 18 species are recorded in North Carolina. For accurate separation of the species one must refer to the technical manuals.

146. Meadow-rue *(Thalictrum).* Perennials 2-11 ft. high; flowers not showy, in loose terminal clusters; stamens and carpels frequently borne in separate flowers; sepals small usually greenish, falling early; petals absent: leaves compound with the rounded, coarsely toothed or lobed leaflets borne on slender petioles of varying length. Six species, one of the tallest of which occurs in marshes, the others in upland woods, especially frequent in the mountains. Spring and summer.

Water-shield Family *(Cabombaceae)*

147. Carolina water-shield *(Cabomba Caroliniana)*. Perennial with stems coated with gelatinous material; few flowered, the flowers borne singly from upper leaf axils; sepals 3 petal-like, petals 3 with lobes on the base; stamens 6: under water leaves thread-form, much fork-divided; floating leaves with narrow blades attached to petiole at center. In quiet water, eastern. Spring and summer.

148. Frog-leaf *(Brasenia purpurea)*. Perennial with flowers on long stalks with 3 sepals and 3 petals, purplish; stamens many, carpels 4-8, separate: leaf-blades all floating, broadly elliptic attached to petiole at center. In quiet water, eastern. Summer.

Lotus Family *(Nelumbonaceae)*

149. American lotus *(Nelumbo lutea)*. Coarse aquatic perennial with the thick creeping stems at the bottom sending up flower stalks bearing large cream-colored flowers (4-10 in. dia.); sepals and petals both colored; stamens numerous; carpels borne in an inverted cone-shaped receptacle prominent in the center of the flower: leaf-blades circular in outline attached to the stalk at the center, not split on one side as in the water lily. In shallow water, not common in our area. Summer.

Water Lily Family *(Nymphaeaceae)*

150. Spatterdock *(Nymphaea)*. Aquatic perennials with the flowers borne above water surface; sepals 5-6, yellow; pistil composed of many carpels combined, the compound stigma flattened above showing number of carpels by radial division lines: leaf-blades with prominent basal lobes. Stream borders and ponds. Summer.

KEY TO SPECIES

Leaf-blades oval in outline, eastern and western.
Northern spatterdock *(N. advena)*.

FIG. 163. LIKE GEMS RESTING ON DARK VELVET, THE WATER-LILIES ARE SHOWN IN AN ESPECIALLY RICH SETTING, FLOATING ON THE DARK WATERS OF AN EASTERN POND. PHOTO BY H. L. BLOMQUIST.

Leaf-blades arrow-shaped, eastern only.
Southern spatterdock (*N. sagittaefolia*, fig. 27.)

151. Water lily (*Castalia odorata*, fig. 163). Familiar perennial water plants with floating sweet-scented flowers of great beauty; stamens grading into petals by expansion of stalk; compound pistil with wheel-like stigma showing radiating lines: leaf-blades broadly oval or orbicular, deeply cleft on one side, the basal lobes point-tipped. Quiet water, central and eastern. Summer.

MAY APPLE FAMILY (*Podophyllaceae*)

152. May apple (*Podophyllum peltatum*). Perennial, having but one flower to a plant, borne from the axil between the two large leaves with deeply lobed blades; sepals 6, petal-like, petals 6-9; stamens many; carpel 1: leaf margin with shallow teeth. Occurs in local societies of varying size. Woods. Spring.

The quaint description of this plant by John Brickell, early North Carolina naturalist, will be of interest here.

"The May Apple so called from its having Apples in the

FIG. 164. THE UMBRELLA LEAF IS A VERY FITTING PLANT FOR THE SHOWERY HIGH MOUNTAINS. PHOTO BY A. F. ROLLER.

month of May; grows upon one Stalk about one-half a Foot high, and has leaves very near as large as a man's Hand, underneath which grow one Apple on each stalk, about the bigness of a musket ball."

153. Twin-leaf *(Jeffersonia diphylla).* Perennial, 6-8 in. high, with the flower stalk from ground; sepals 4, petal-like, petals 8, white; stamens 8; stigma 2 lobed: leaf-blades divided by 2 deep clefts from opposite sides forming twin-halves, each with a pointed tip and rounded base. Mountain woods. Spring.

154. Umbrella-leaf (*Diphylleia cymosa,* fig. 164). Coarse perennial, 1-3 ft. tall, with white flowers in a terminal umbellate cluster; sepal, petals and stamens each 6 in number; pistil 1: basal leaves large, blades round in outline but deeply lobed; stem leaves 2, blades lobed. In very wet ground, mountain slopes. Spring.

155. Blue cohosh *(Caulophyllum thalictroides).* Perennial, 1-3 ft. high, with many greenish purple flowers in a rounded loose terminal inflorescence; sepals 6 not counting the

FIG. 165. THESE BLOODROOTS HAVE BECOME HIGHLY SOCIAL.
PHOTO BY MISS SUSAN IDEN.

little bracts near them, petals 6, short, stamens 6, pistils 1: leaves compound, the leaflets commonly borne in threes; each leaflet 3-lobed at end; leaves basal except 2 on stem. Mountain woods. Spring.

BLOODROOT FAMILY *(Papaveraceae)*

156. Bloodroot (*Sanguinaria Canadensis,* fig. 165). Perennial with white showy flower borne directly on underground stem; sepals 2, petals 8-10, early shedding; stamens many, carpels 2 united in single pistil with 2 stigmas: leaf-blades round or kidney-shaped in outline, lobed. Plant with reddish juice throughout. Upland woods. Spring.

157. Mexican poppy *(Argemone Mexicana).* Annual or biennial weed, 1-2 ft. tall, with large (1-2 in. broad) showy yellow or cream-colored flowers borne from the upper leaf axils; stamens numerous: leaf-blades irregularly lobed and cut, very spiny or thistle-like, alternate. Occasional in waste places. Adventive from tropical America. Summer.

FUMITORY FAMILY *(Fumariaceae)*

158. Allegheny vine *(Adlumia fungosa).* Plant vine-like: flower pink or white, nearly 3/4 in. long, with the 4 petals united, the corolla tube forming a flattened slender heart-shaped structure, tipped by the slightly spreading petal ends (resembles the bleeding heart of the gardens): leaves compound, the thin, delicate leaflets borne on slender stalks. Mountain woods. Summer.

159. Dutchman's breeches and related species (*Bicuculla*, fig. 160). Perennials with white flowers similar to the preceding but the four petals in 2 pairs not completely united, merely lightly coherent: leaf-blades finely divided and lobed. Mountain woods. Spring.

KEY TO SPECIES

Flowers in a simple raceme, white.
  Basal lobes of flower well developed, spreading apart.
    Dutchman's breeches *(B. Cucullaria).*
  Basal lobes not prominent, the base simply heart-shaped.
    Squirrel corn *(B. Canadensis).*
Flowers in a branching raceme, pink.
    Wild bleeding heart *(B. eximia).*

160. Roman wormwood and colic weed *(Capnoides).* Annual or biennial with the flowers in racemes, bilateral; sepals 2, petals 4, the exterior 2 unlike, one short-spurred; stamens 6; leaves compound, the leaflets deeply lobed. Mountain woods chiefly. Summer.

KEY TO SPECIES

Flowers purple or pink, plant 1-3 ft. high.
    Purple roman wormwood *(C. sempervirens).*
Flowers yellow, plant 4-12 in. high. Colic weed *(C. micranthum).*

MUSTARD FAMILY *(Brassicaceae)*

161. Yellow cress *(Barbarea).* Biennials or perennials, 1-2 ft. high, with loose terminal racemes of yellow flowers; pod long and narrow, slightly 4-angled: leaf-blades pinnately divided, the terminal lobe much the largest and rounded at the tip, smooth, alternate. Weeds from Europe widely spread in waste places.

162. Mustard *(Brassica and Sinapis).* Mostly annual weeds with rather showy yellow flowers in a loose terminal cluster; pods slender: leaves simple or compound. In waste places; introduced from Europe. Summer and fall.

KEY TO COMMON SPECIES

Pods constricted between the seeds.
- Leaves compound with large terminal lobe. Charlock or white mustard *(Sinapis alba).*
- Leaves simple, tooth or lobed. Wild mustard *(Brassica arvensis).*

Pods not constricted between the seeds.
- Upper leaf-blades not clasping, merely sessile. Black mustard *(B. nigra).*
- Upper leaf-blades clasping. Wild turnip *(B. campestris).*

163. Water cress *(Roripa).* Perennials with small flowers (petals under 1/4 in. long): above water leaf-blades tapering toward base, pinnately divided: pods slender-cylindric to oval or nearly globular, in small streams. Summer.

KEY TO SPECIES

Flowers white.
- Leaves all alike, not a native, from Europe. True water-cress *(R. Nasturtium).*
- Leaves under water different from those above. Wild water-cress *(R. Americana).*

Flowers yellow. Three native species much alike.

164. Wormseed mustard *(Erysimum cheiranthoides).* Perennials, 1-2½ ft. with long narrow racemes of yellow flowers; petals about 3/8 in. long; pod linear, 1/2-3/4 in. long: leaves narrow, nearly linear, 1-4 in. long. Dry uplands. Summer.

165. Sea rocket *(Cakile).* Annual, 1-1½ ft. high with flowers in simple racemes; pods 2-jointed, each joint with one seed: leaf-blades fleshy, coarsely toothed. Plants confined to coastal sand near the sea. Not to be confused with sea elder *(Iva imbricata)* which it superficially resembles. The fruit of the latter is but single jointed. On coastal dunes. Summer.

166. Whitlow-grass *(Draba).* Annuals, winter annuals or perennials, 1-7 in. high with simple racemes of tiny white flowers arising from a basal rosette or tuft of leaves; length of pod not over 6 or 7 times its width; commonly shorter: leaf-blades oval or nearly so. Five species of unimportant minute weeds. In sunny upland meadows. Spring.

167. Toothwort *(Dentaria).* Perennials, 8-15 in. high, with purple or white flowers in a terminal raceme; sepals about 3/8 in. long; flowers white or purple; pod linear; leaves chiefly basal, blades palmately divided into 3 parts, the outer two sometimes divided again into 2 lobes; margin prominently toothed. Four species of upland moist woods, central and western. Spring.

168. Bitter cress *(Cardamine).* Annuals, biennials or perennials with white, pink or purple flowers in racemes, petals contracted below into stalks; pod linear, flat, erect: two species have simple leaves, the other 6 have pinnately divided leaves distributed chiefly on the stem. Species of moist ground or damp woods, chiefly western. Spring and summer.

169. Rock cress *(Arabis).* Annuals, biennials or perennials with white flowers in our species; sepals not over 3/8 in. long; pods narrow, flat: leaves mostly simple (pinnately divided in *A. Virginica*) sessile on stems or tufted at base. Rocky or dry soil, central and western. Spring.

170. Tansy mustard *(Sophia pinnata).* Perennial, 8-24 in. tall with yellow flowers in terminal racemes, which in fruit become elongated; pods short not over 4 times as long as wide; seeds in 2 rows: leaf-blades finely divided, the primary pin-

Fig. 166. The sun-dew brightens the black bog soil with its red rosettes of sticky leaves.

nate divisions being divided again. Dry uplands. Spring and summer.

Caper Family *(Capparidaceae)*

171. Spider-flower *(Cleome spinosa)*. Annual weed, 2-4 ft. high with peculiar purplish flowers having long exserted stamens, borne in a terminal leafy-bracted raceme; flowers slender-stalked, petal blades stalked; leaves below palmately compound with 5-7 leaflets changing above into the simple bracts, all alternate; stems and leaves sticky-pubescent. In waste places especially in our western area. Adventive from tropical America. Summer.

Sundew Family *(Droseraceae)*

172. Sundew (*Drosera,* fig. 166). Biennial or perennial small bog plants with rosettes of reddish leaves, bearing curious sticky upright or curved structures which catch small insects: the tiny white or pink flowers are in a simple raceme,

FIG. 167. THE FLY-TRAPS BLOOM IN LATE MAY, WHEN THEY MAY MOST READILY BE FOUND WITH THE AID OF THE FLOWERS SHOWING ABOVE THE SURROUNDING GRASS.

the number of sepals, petals and stamens, varying from 4-8: the leaf-blades vary in shape in the different species, the commonest one being the round-leaved sundew; the other 3 species have more elongated leaf-blades. Four species on peat or the black soil of bogs, eastern. Spring and summer.

173. Venus' fly-trap (*Dionea muscipula,* figs. 67, 69, 167). Perennial with flowering stalk 4-12 in. high, bearing the white flowers in an umbel at the apex; flower 5-parted except stamens which are commonly 10 in number, sometimes more; petals delicately veined: the bristle-fringed leaf-blades, marvelously modified into an efficient insect trap, the petioles merely being expanded into flat leaf-like organs. For a fuller account see descriptive chapter on the savannahs. In sandy bogs, eastern. Early summer.

174. Pitcher plants and trumpets (*Sarracenia*). Perennials with nodding large flowers with enormously expanded stigmas resembling an umbrella; 5-parted with thick sepals and prominent drooping petals: leaf petioles modified with water retaining vases, topped at on side by the smaller leaf-blade. For fuller account of these insect consuming plants see chapter on the savannahs. In grass-sedge bogs or savannahs, eastern and central. Spring.

KEY TO SPECIES

Leaves lying on side, curving upward toward end.
Pitcher-plant *(S. purpurea)*.
Leaves erect.
Blade portion or hood fitting closely down over top of vase.
Rain-hat trumpet (*S. minor,* fig. 64).
Blade or hood short stalked and borne above the vase top.
Plants under 10 in. high.
Small umbrella trumpet (*S. rubra,* fig. 168).
Plants over 15 in. high.
Large umbrella trumpet (*S. flava,* figs. 63, 169).

River-weed Family *(Podostemaceae)*

175. River-weed *(Podostemon ceratophyllum)*. Aquatic with small flowers enclosed in a bract: stems not over 8 in. long, attached by flattened, basal portion (disk-like): leaf-blades divided into narrow segments, rather rigid. In running water. Summer.

Stone-crop Family *(Sedaceae)*

176. Southern rock orpine *(Diamorpha cymosa)*. Low annuals or biennials, 2-4 in. high with small pink or purple flowers borne in flat-topped clusters; carpels 4 or 5 united below the middle: leaves fleshy short, the shape of capsules. In thin soil on granite rocks, central and western. Spring.

177. Roseroot *(Rhodiola Roanensis)*. Perennial, not over 10 in. tall with dull purplish flowers in flat-topped inflorescence; carpels separate standing erect in the flower: leaf-blades narrow, 1/2-1½ in. long, flat. Only reported on Roan Mountain cliffs. Spring and summer.

FIG. 168. THE LITTLE RED TRUMPETS *(Sarracenia rubra)* ARE AN INTERESTING BUT NOT COMMON ADDITION TO OUR INSECTIVOROUS FLORA.

FIG. 169. THE TRUMPET FLOWER STIGMA (LEFT) IS NOTED FOR ITS SIZE AND THE PETALS (RIGHT) FOR THEIR VIOLIN-LIKE SHAPE.

178. Four-fruited stone crop *(Tetrorum pusillum).* Low annual, 2-5 in. high, spreading, with small white or pink star-like flowers in flat-topped clusters; carpels 4, not erect but spreading: leaves short about 3/8 in. long fleshy. Chiefly on rocks in the mountains. Spring.

179. Stonecrop *(Sedum).* Low and spreading annuals or perennials with star-like flowers in flat-topped clusters; carpels as many as the petals, spreading; stamens twice as many: leaves fleshy varying from cylindric to flattened types. Five species chiefly on rocks in the mountains. Spring.

KEY TO SPECIES

Petals white.
  Lower leaves alternate. — Nevius' stonecrop *(S. Nevii).*
  Lower leaves in threes. — Common wild stonecrop *(S. ternatum).*
Petals yellow. — Wall pepper *(S. acre).*
Petals purple. — Widow's cross *(S. pulchellum).*
Petals pink. — American orpine *(S. telephoides).*

DITCH STONECROP FAMILY *(Penthoraceae)*

180. Ditch stonecrop *(Penthorum sedoides).* Perennial 4-30 in. high, with dull white flowers in flat-topped clusters; flowers without petals, but the simple ovate sepals petal-like; carpels 5 or 6, united below, the group contracted at the middle: leaves simple, alternate, blades oval or elliptical with rather fine teeth. In wet soil. Summer.

GRASS OF PARNASSUS FAMILY *(Parnassiaceae)*

181. Grass of Parnassus *(Parnassia).* Perennials with flowers solitary on slender stalks near the middle of which a leaf is frequently borne; petals with conspicuous nerves; false stamens without anthers occur between the 5 true stamens: leaf-blades oval or round, thickish, leaves all basal except the one (bract-leaf) on the stem. In wet areas, chiefly in the mountains. Summer.

SAXIFRAGE FAMILY *(Saxifragaceae)*

182. Golden saxifrage *(Chrysosplenium Americanum).* Spreading semi-aquatic perennial with the greenish yellow flowers, solitary, sepals 4 petal-like; petals absent; styles 2;

FIG. 170. FOR MOIST ROCK GARDENS NOTHING CAN SURPASS THE FOAM FLOWER.

leaf-blades (not over 3/4 in. dia.) nearly round in outline or kidney-shaped with the margin shallowly lobed. In mountain brooks or cool shady wet soil, western. Spring.

183. Bishop's Cap *(Mitella diphylla)*. Erect perennial, 10-18 in. tall, with white flowers in terminal raceme; petals deeply divided (an unusual character) or cut: leaves all basal except a pair on the stem; leaf-blades heart-shaped or tending to be 3 lobed; margin with coarse, shallow teeth. In moist woods, chiefly in the mountains. Spring.

184. Alum-Root *(Heuchera)*. Perennials, 1-3 ft. high, with small unattractive greenish or purplish flowers in a very loose terminal cluster, borne commonly on a long slender stalk; sepals frequently unequal and attached at base to the ovary: leaves mostly basal, long stalked with the rounded shallowly lobed blades, as well as the petioles prominently hairy; margin coarsely toothed. About eight species, in upland woods, chiefly central and western. Spring and summer.

FIG. 171. ONE OF THE EARLIEST WILD FLOWERS OF THE PIEDMONT AND MOUNTAIN REGIONS IS THE COMMON SAXIFRAGE. PHOTO BY H. L. BLOMQUIST.

185. Foam flower (*Tiarella cordifolia,* fig. 170). A rather delicate perennial, 6-12 in. tall with small white flowers in a conical raceme; calyx practically free from the ovary; the 2 carpels are unequal: leaves basal much like those of the preceding genus. Moist woods, central and western. Spring.

186. Aconite saxifrage *(Therophon aconitifolium).* Stout perennial, 1-2 ft. high, with small inconspicuous white flowers in a loose terminal cluster, the main branches of which arise from the axils of conspicuous leaf-like bracts; sepals united, the tube attached to the ovary: leaves chiefly on the stem, blades kidney-shape in outline (2-7 in. wide), the margin lobed and sharply toothed. In mountain woods. Summer.

187. Saxifrage *(Micranthes).* Low perennials, not over 12 in. high with small white flowers in open terminal clusters borne on slender stalks; carpels united below and slightly united to the calyx tube: leaves in a basal rosette, blades varying from round to oval and in the mountain lettuce *(M. micranthidifolia)* to narrow oblanceolate; margin toothed. Our commonest species is the early saxifrage (*M. Virginiensis,* fig.

171) with oval short-stalked leaves with no yellow markings on the petals. Four species in upland woods. Spring.

188. High mountain saxifrage *(Hydatica petiolaris).* Perennial with white flowers in a loose terminal cluster reaching as high as 20 in. from ground; petals contracted at base, the 3 upper broader than the 2 lower: leaf-blades narrow tapering into winged petioles, margin with very coarse sharp teeth. On exposed rocky places on mountain summits. Summer.

189. False goat's-beard *(Astilbe).* Tall showy perennials, 2-6 ft. high with numerous small white flowers arranged thickly along many raceme-like branches which extend from the main branches of the compound inflorescence; the whole inflorescence may be as much as a foot in length: leaves compound with the toothed leaflets in threes. Two species much alike of which *A. biternata* is the common one, the other *(A. crenatilobata)* only being reported from Roan Mountain. Deep mountain woods. Summer.

ROSE FAMILY *(Rosaceae)*

190. True goat's-beard *(Aruncus aruncus).* Tall perennial, 3-7 ft. high, resembling the preceding genus from which it may be readily distinguished by the numerous stamens (more than 10) in the small flowers; carpels 3; petals oval: leaves compound, leaflets in threes or fives at end of branches, toothed. In mountain woods. Summer.

191. Indian physic *(Porteranthus).* Perennials, 2-4 ft. high, with white flowers in a loose terminal cluster; petals narrow, slightly unequal; carpels 5: leaves compound, leaflets 3, oval or narrower with tapering tips, toothed; one species has paired stipules which resemble small leaves. Two species in upland woods of the piedmont and mountains. Summer.

192. Three-toothed cinque-foil (*Sibbaldiopsis tridentata,* fig. 172). Low woody trailing perennial with white flowers in a flat-topped cluster; 5 narrow bractlets occur on the calyx

FIG. 172. THE THREE-TOOTHED CINQUEFOIL MAY BE FOUND ONLY ON MOUNTAIN SUMMITS. PHOTO BY H. L. BLOMQUIST.

tube in addition to the broader sepals: leaflets 3, each 3 toothed at the blunt end. On exposed rocks of mountain tops. Summer.

193. Wild strawberry *(Fragaria).* Low trailing perennials with white flowers in clusters of 3-6; petals orbicular, contracted below with short stalks; petals and stamens borne on calyx tube (hypanthium): leaf-blades of 3 leaflets, sessile, coarsely toothed: fruit similar to cultivated strawberry but not so large. Two species of dry uplands. Spring.

194. Indian strawberry *(Duchesnea Indica).* A perennial resembling the preceding except that the saucer-shaped part of the calyx (hypanthium) bears prominent leaf-like appendages between the sepals: the fruit, though red and appearing like a strawberry, is dry and inedible. An escaped plant whose native home is India. Summer.

195. Cinquefoil or five finger *(Potentilla).* Low or medium perennials with yellow flowers borne separately or in clusters: stamens many, about 20; carpels many: leaflets 3 or 5, toothed. Dry uplands. Spring.

KEY TO COMMON SPECIES

Leaflets 5, plant trailing.
 Finely pubescent. Common five finger *(P. canadensis)*.
 Coarsely pubescent. Dwarf five finger *(P. pumila)*.
Leaflets 3, plant erect, 1-2½ ft. tall.
 Rough five finger *(P. Monspeliensis)*.

196. Barren strawberry *(Waldsteinia)*. Perennials which resemble the wild strawberry but the flowers are yellow and produce no fleshy fruit; carpels 2-6: leaflets roughly triangular in outline, coarsely toothed, especially toward the tip. Two species in shady woods. Spring and summer.

197. Avens *(Geum)*. Erect pubescent perennials, 1-2½ ft. high, with few yellow flowers borne in open cluster above; carpels many forming ovoid head in center of flower, the style with the very characteristic bent portion, suggesting a bayonet in shape: leaflets 3 or more variously toothed and cut. In moist woods, central and western. Summer.

198. Mountain avens *(Sieversia radiata)*. Perennial, 6-24 in. tall, with showy yellow flowers solitary or but a few in the cluster; petal tips deeply notched (heart-shape); corolla nearly 1½ in. dia.; basal leaves pinnately divided with long petioles the stem leaves sessile, cleft into many segments. On summits of high mountains. Summer.

199. Agrimony *(Agrimonia)*. Rather coarse perennials, 1-5 ft. high with yellow flowers in narrow racemes; calyx tube prickly with hooked bristles; carpels 2 enclosed by calyx tube: leaves pinnately compound, the leaf stem bearing small leaflets between the larger ones. Four or 5 species much alike. Woods and openings. Summer.

200. American burnet *(Sanguisorba Canadensis.)* Tall smooth perennials, 1-6 ft. high with white flowers in a terminal showy spike; the 4-winged calyx tube contracted at the top, the 4 lobes (sepals) petal-like; petals absent: leaves pinnate; leaflets toothed; stipules leaf-like. In moist ground and edges of marshes, western. Summer.

MIMOSA FAMILY *(Mimosaceae)*

201. Sensitive brier *(Morongia).* Curious trailing very prickly perennials with small purplish flowers in a showy spherical head (3/4-1 in. dia.); petals united up to the middle of flower: leaves twice pinnate with very small sensitive leaflets which become pressed together when the leaf is struck. In dry thickets or open places. Summer.

SENNA FAMILY *(Cassiaceae)*

202. Senna *(Cassia).* Medium to tall annuals or perennials, 2-7 ft. high with the few yellow flowers borne in axillary and terminal clusters; petals nearly alike; stamens 10; leaves alternate, evenly pinnate; leaflets not toothed. Commonly in moist soil. Summer.

KEY TO SPECIES

Leaflets few (4-6), pod long, very slender and curved like a sickle.
Sickle senna *(C. Tora).*

Leaflets 8 or more.
Leaflets lanceolate, broadest below middle.
Coffee senna *(C. occidentalis).*
Leaflets elliptical, broadest about middle.
American senna *(C. Marylandica).*

203. Sensitive pea *(Chamaecrista nictitans).* Low and spreading annual with inconspicuous yellow flowers borne singly above the leaf axils; leaflets 12-44 sensitive to touch, i. e., closing together in pairs: leaflets narrow, small, under 1/2 in. in length. Note curious gland near base of the leaf stalk. Open woods and fields. Summer.

PEA FAMILY *(Leguminosae)*

204. Bush pea *(Thermopsis).* Perennials, 1-3 ft. high, with yellow flowers mostly terminal in racemes; stamens 10, all distinct; ovary sessile; pod linear: leaves trifoliate with leaf-like stipules. Four species of open or dry woods of mountain slopes. Spring and summer.

FIG. 173. THE SANDHILL LUPINE IS A FLOWER PRODUCER OF THE FIRST RANK. PHOTO BY H. L. BLOMQUIST.

205. **Wild indigo** *(Baptisia)*. Perennials similar to the preceding except the ovary is stalked and the stipules are not so prominent: leaves trifoliate and in the common wild indigo *(B. tinctoria)* seldom over 1 in. long. In our other species they are definitely larger. They turn blue-black upon drying, hence the common name. Six species in the drier areas of open broad-leaved woods, pine forest, and sandhills. Summer.

206. **Rattle box** *(Crotalaria)*. Annual or perennial, 5-20 in. high, somewhat spreading plants with the few yellow flowers in racemes; anthers of 2 kinds, alternating; mature fruit pod inflated with the loose seeds rattling within when shaken: leaves apparently simple (1 leaflet) with prominent pointed stipules suggesting the base of an arrow in shape. In *C. sagittalis,* these stipules are strongly decurrent or extend downward on the stem. Two species in sandy soil, chiefly eastern.

207. **Lupine** *(Lupinus)*. Attractive spreading perennials with showy racemes of blue flowers; calyx prominently 2-

lipped; leaves of *L. perennis* are palmately compound with 7-11 leaflets; the leaves of *L. diffusus* (Fig. 173) found in the sandhills are simple consisting of one elliptical, finely hairy leaflet. Upland woodlands. Spring.

208. **Clover** *(Trifolium).* Annuals or perennials with small white purplish or yellow flowers mostly in compact rounded or elongated heads; petals generally remaining on flowers after drying: leaves of 3 (rarely more) leaflets, slightly toothed or entire. Mostly introduced species on uplands. Spring and summer.

KEY TO THE COMMON CLOVERS

Flowers white.
  Leaflets narrow, 3 times as long as wide. Rabbit-foot clover *(T. arvense).*
  Leaflets broad, nearly circular in outline. White clover *(T. repens).*
Flowers yellow.
  All leaflets sessile. Yellow clover *(T. agrarium).*
  Central or terminal leaflet stalked.
    Heads few-flowered, under 15. Least hop-clover *(T. dubium).*
    Heads with numerous flowers, over 20. Hop-clover *(T. procumbens).*
Flowers bright red. Crimson clover *(T. incarnatum).*
Flowers purplish.
  Individual flowers not stalked and heads with leaves from the base. Red clover *(T. pratense).*
  Individual flowers short stalked and heads on stalks above nearest leaves.
    Calyx about as long as the corolla. Carolina clover *(T. Carolinianum).*
    Calyx about ½ as long as the corolla or less.
      Heads nearly 1 in. wide. Buffalo clover *(T. reflexum).*
      Heads less than ¾ in. wide. Alsike clover *(T. hybridum).*

209. **Bird's-foot trefoil** *(Lotus Helleri).* Annual, 8-24 in. high, the erect stems bearing small (1/4 in. long) pink flowers in the axils of the upper leaves; pods straight, about 1 in. long: leaflets 3, except upper ones, the blades narrow, nearly linear. In sandy pine woods, central. Summer.

210. Goat's-rue *(Cracca).* Perennials, 1-3 ft. high, with slender stems bearing the flowers in rather showy racemes; flowers of the common *C. Virginiana* are predominantly yellowish white, with purplish tinged parts; those of the less common *C. hispidula* are white at first but turn pink and finally red in age: leaves pinnate with many elliptical, more or less hairy leaflets. In sandy soils, chiefly eastern and central. Summer.

211. Pea tree *(Sesban macrocarpa).* Very tall annual, 4-12 ft. with few yellowish purple spotted flowers in racemes: leaves with 12-35 pairs of sessile elliptical leaflets: pod very slender, jointed. In moist places. Spring to fall.

212. Giant pea *(Glottidium vesicarium).* Annual, coarse herb, 3-12 ft. tall with spreading branches bearing purple tinged, yellow pea-like flowers in axillary racemes: leaves equally pinnate, leaflet number averaging around 40: pods stalked. In moist low ground, chiefly eastern. Summer.

213. Yellow goat's-rue *(Tium apilosum).* Slender perennial, 1-3 ft. high, stems bearing the few flowers in a raceme; flowers yellow, not tinged with purple; pod narrow tapering at ends, in cross section appearing as if 2-celled by a deep intrusion of the wall along the upper and under sides: leaves pinnate; leaflets 15-25, each with a short stalked narrow blade. In coarse sand areas, eastern. Spring and early summer.

214. Blue darts *(Psoralea).* Perennials, 1-2½ ft. high with showy slender pointed racemes of many blue or purplish flowers; pod not splitting open: leaves palmately compound. Chiefly in sandy soils, eastern and central. Spring and summer.

KEY TO SPECIES

Leaflets 5-11, very narrow, almost thread-like. *(P. Lupinellus.)*
Leaflets 3.
  Corolla over ⅜ in. long. *(P. canescens).*
  Corolla under ⅜ in.
    Leaflets less than ¾ in. broad. *(P. pedunculata).*

Leaflets over ¾ in. broad.
Leaflets with rounded base. *(P. Onobrychis).*
Leaflets with heart-shaped base. *(P. macrophylla).*

215. Lead plant *(Amorpha herbacea).* Low shrub-like plant with compact spike-like racemes of small, blue flowers under 1/4 in. long: leaves unequally pinnate, leaflets numerous, elliptic, together with the stems covered by a white pubescence: pods under 1/4 in. long. In dry, sandy soils, eastern. Spring.

216. Summer farewell *(Kuhnistera pinnata).* Perennial, 1-2 ft. tall, the stems in clumps bearing compact clusters or heads of small white flowers; flower unlike most legumes has only 5 stamens with united filaments, the four lower petals are alike, contracted below into stalks which are united to the stamen tube: leaves compound, the leaflets very slender, thread-like. In coarse sandy soils, eastern. Fall.

217. Joint vetch *(Aeschynomene Virginica).* Tall weed-like perennial, 2-6 ft. high with reddish tinged flowers 3/8 in. long borne in clusters at the end of the inflorescence stalk; pod with 6-10 square joints: leaves pinnate, the narrow leaflets under 1/2 in. in length, numerous. In moist sandy soil, chiefly in the coastal plain. Summer and fall.

218. Hide and seek *(Zornia bracteata).* Spreading herbs with yellow flowers in rather compact clusters; each flower enclosed and nearly hidden by a pair of leaf-like bracts: leaves compound, leaflets 4, oval. Sandy soil, eastern. Spring and summer.

219. Pencil flower *(Stylosanthes).* Small erect or reclining plants, 8-20 in. tall, with few yellow flowers: calyx in form of slender tube with 5 lobes, the 4 upper united to form a lip opposite to the lower; petals and stamens attached near summit of the calyx tube: leaflets 3 from a short petiole, which has stipules at the base which sheath the stem. Two species: *S. biflora* with the blade of the large petal longer than broad and *S. riparia* with the blade broader than long. In dry woods. Spring to fall.

220. **Tick-trefoil** *(Meibomia).* A large genus of low to tall herbs, 4 in.-6 ft. high, with purplish flowers in open racemes or loose clusters; fruit flat, jointed with constrictions between the joints: leaflets 3, with leaflet stipules (stipels) at the base of each.

There are close to 20 species of tick-trefoils or beggar ticks in the state all of which may be recognized by the characteristic angular fruit joints which adhere so tightly to clothing by their flat faces. In woods and thickets. Summer and fall.

221. **Bush clover** *(Lespedeza).* Trailing or erect herbs with violet purple flowers in loose or compact clusters; pod mostly of one joint, tipped with a point: leaflets 3, leaflet stipules not present. Two native species *(L. repens* and *procumbens)* have a trailing habit: the other 8 or 9 are erect and generally assume a slender wand-like aspect with the flowers in close clusters on the upper part of the stem. Dry upland soils of woods and open places. Summer and fall.

222. **Hairy-fruited bean** *(Dolicholus).* Perennials, 4-12 in. high, with yellow flowers in short racemes; seeds commonly 2 to the pod: leaves simple, round in outline in *D. simplicifolius,* compound with 3 oval leaflets in the other two species; commonly dotted with resinous particles. In dry soils and sandhills. Summer.

223. **Scarlet-flowered bean** *(Erythrina herbacea).* Low and spreading perennial with scarlet flowers in racemes; flowers large (1½ in. long); pods contracted between the seeds: leaflets 3, triangular in outline, 1½-3½ in. long. Sandhills. Spring.

224. **Milk pea** *(Galactia).* Trailing or vine-like perennials, with purplish flower-clusters on the end of axillary inflorescence stalks; calyx lobes unequal: leaflets 3, oval or elliptical, never over 2 in. in length. Four or five species,

variable. Chiefly in coarse sandy soils, eastern. Spring and summer.

225. **Hog-peanut** *(Falcata comosa).* Twining perennial with purple or white flowers in axillary racemes; calyx tubular, upper lobes shorter than lower; large petal somewhat folded about other petals at base: leaflets 3, oval with tapering pointed tips, 1-3 in. long. Damp woodlands, central and eastern. Summer and fall.

226. **Virginia butterfly pea** *(Bradburya Virginiana).* Twining with showy light purplish flowers much resembling the commoner butterfly pea described below: upper calyx lobe twice as long as the calyx tube; pod linear; blade of large petal typically broader than high: leaflets 3, narrowly-oval, 3/4-3¾ in. long. In dry or sandy soil. Summer.

227. **Butterfly pea** *(Clitoria Mariana).* Trailing or climbing with showy lilac-blue flowers borne singly or in clusters of 2 or 3. Calyx tubular, the two upper teeth partly united; pod elliptical: blade of large petal definitely longer than broad: leaflets 3, ovate-lanceolate. In dry woods and openings. Spring and summer.

228. **Bean-vine** *(Phaseolus polystachyus).* Annual twining stems from a perennial root bearing racemes of purplish flowers; style within the curved keel (2 lower and united petals) bearded along the inner side; pod curved slightly, tapering toward the base: leaflet-blades 3, broadly ovate the middle one stalked, the lateral ones, sessile or nearly so. In open woods and thickets, more frequent westward. Summer.

229. **Wild bean** *(Strophostyles helvola).* Trailing with purplish flower clusters on the end of long stalks; flowers greenish when faded: leaflets 3, each frequently more or less 3 lobed 3/4-1½ in. long. In dry woodlands, preferring sandy soils. Summer and fall.

230. **Vetch** *(Vicia).* Common trailing and climbing herbaceous vines with few or many blue, purple, or white flowers in racemes; upper calyx lobes shorter than lower ones;

large petal notched at tip: leaves pinnate, terminating in a tendril; leaflets many (6-12), commonly narrow. Six species most of which are foreigners. The native *V. Caroliniana* is an attractive species with its many-flowered racemes of light blue flowers. In dry woodlands. Spring and summer.

231. **Wild sweet pea** *(Lathyrus).* Climbing or trailing with rather showy purple flowers in clusters of 2 or more; calyx tube somewhat inflated on one side at base; large petal notched at tip and contracted below to a short stalk: leaves terminating in a tendril: leaflets variable, 2 in *L. pusillus,* 4-8 in *L. myrtifolius* and 8-14 in *L. venosus.* In moist soil except *L. pusillus.* Spring and summer.

Crane's-bill Family *(Geraniaceae)*

232. **Crane's-bill** *(Geranium).* Perennials, 1-2 ft. high with few purple flowers in a terminal cluster; styles 5, separate; leaf-blades palmately lobed, the lobes coarsely toothed. Spring and summer.

KEY TO SPECIES

Flowers 2 or 3 in cluster, 1 in. or more in dia., plants of sandy woods. True crane's bill *(G. maculatum).*

Flowers many in flat-topped cluster, ½ in. dia., plant of open dry areas. *(G. Carolinianum).*

Jewel-weed Family *(Balsaminaceae)*

233. **Jewel-weed** *(Impatiens).* Frail annuals, 2-5 ft. high with watery stems, swollen at the nodes; flowers showy, pendent; stamens with short filaments; lateral petal pairs united: leaf-blades with wavy margins, petioled, 1-5 in. long. In very moist soil. Summer.

KEY TO SPECIES

Flowers orange with a long spur ½ as long as corolla. *(I. biflora).*

Flowers yellow with a short spur ¼ as long as corolla, western. *(I. pallida).*

Flax Family *(Linaceae)*

234. **Wild flax** *(Carthartolinum,* fig. 174). Annuals or perennials, 8 in.-2½ ft. high, with yellow flowers in loose

FIG. 174. THE WILD FLAX IS FREQUENTLY SEEN IN BOGGY AREAS.

terminal clusters; carpels 5 wholly or partly united: leaves simple, narrow, alternate or opposite, not over 1 in. long. Five species much alike. In dry soil or savannahs. Summer.

WOOD-SORREL FAMILY *(Oxalidaceae)*

235. Mountain wood-sorrel *(Oxalis Acetosella)*. Perennial, 2-6 in. high with white or pink striped flowers solitary on slender stalks from the elongated scaly underground stems: leaflets 3, the blades broader than they are long. Cool damp woods of the mountain tops, especially abundant in the moss *(Hylocomium)* beds of the balsam-fir forest, western. Summer.

236. Violet wood-sorrel *(Ionoxalis violacea)*. Perennial, 4-9 in. high, with rose-purple flowers in umbels of 3-10, the main stalk arising from a scaly bulb-like underground stem: leaflets 3, about as broad as long, each with a small orange-colored gland in the notch. In woods, chiefly in the mountains. Summer.

237. Yellow wood-sorrel *(Xanthoxalis).* Annuals or perennials with the yellow flower clusters borne on stems above the ground; fruit capsules slender and commonly borne erect; leaflets 3, sessile on the end of slender petioles which are arranged alternately on the main stem. Ten species much alike. Most of our species prefer moist habitats. Spring and summer.

Milkwort Family *(Polygalaceae)*

238. Milkwort *(Polygala).* Low-medium herbs with very peculiar small flowers arranged in racemes: the flower is not only bilateral but 2 of its sepals are enlarged and colored, imitating petals; the 3 petals are united and at the base of the middle corolla lobe is a scale or scales, accessory structures only to be observed with a good lens; the corolla tube is split on the upper side and more or less united to the 8 stamen stalks: leaves simple, narrow, seldom over an inch long, alternate, opposite or whorled. About 16 species differing in color or other less striking characters. The habitat range is a wide one.

KEY TO SPECIES

Flowers yellow.
  Flowers at first in head-like inflorescence, later much elongating: bog plant.
    Stems under 1½ ft. tall. Red hot poker *(P. lutea).*
    Stems over 1½ ft. Tall yellow milkwort *(P. Cymosa).*
  Flowers not in head-like inflorescence, but a spreading flat-topped one. Branched yellow milkwort. *(P. ramosa).*
Flowers greenish.
  Leaves alternate. Senaca snakeroot *(P. Senega).*
  Leaves whorled. Whorled-leaved milkwort *(P. verticillata).*
Flowers white. No white species are known, the white-flowered plants being exceptional variants of the purple flowered species.
Flowers purple. About a dozen species—the purple milkworts, which need not be separated in a popular treatment.

Spurge Family *(Euphorbiaceae)*

239. Tragia *(Tragia).* A slender perennial, with few erect branches bearing terminal spike-like racemes of small

FIG. 175. TREAD SOFTLY, A FAMILIAR ENEMY WITH A VERY HOT TOUCH DUE TO STINGING HAIRS.

flowers, only the upper staminate ones having the red 4-lobed, corolla-like structure, the lower carpellate or seed-bearing flowers inconspicuous except for the fruit: leaf-blades varying from ovate to linear, the former with a few coarse teeth, sessile or nearly so, alternate. In sandy soils, eastern. Summer.

240. Queen's delight *(Stillingia sylvatica).* Coarse herbs with branches spreading from a common root-crown, 1-3½ ft. tall: staminate flowers each with 2-3 stamens, borne in a yellow spike above the few carpellate flowers, which later develop the 3-lobed fruit. The entire staminate spike is after maturity cut off and dropped; the flowers are without petals: leaves alternate, ovate, tapering below, 1½-4 in. long, shallowly but finely toothed. In sandy soil, especially abundant in the sandhills. Summer.

241. Tread softly (*Cnidoscolus stimulosus,* fig. 175). Perennial, 4 in.-3 ft. high, with white flowers, few in a cluster borne amid the leaves, rather showy; the "corolla" is a 5-lobed calyx; stamens many, 10 or over; borne in separate flowers from the carpellate ones: leaves deeply 3-5 lobed, alternate, thickly covered with the stinging hairs which so successfully defend this plant from man and beast. In dry sandy woods, especially common in the sandhills. Summer.

242. **Spurge** (*Chamaesyce, Tithymalopsis* (Fig. 84) and *Tithymalus*). These genera with about 4 species each in North Carolina are very closely related herbs which need not be distinguished in an elementary manual. They all have extremely minute staminate flowers which are aggregated with one pistillate flower into an inflorescence, which with its 4 or 5-lobed involucre appears as a small individual flower. The involucre commonly bears yellow glands at the base of the lobes. When the lobes are relatively large as in the flowering spurge *(Tithymalopsis corollata)* the plant becomes desirable for cultivation. Most of our spurges show an abundant milky juice in stems and leaves. The western U. S. snow-on-the-mountain *(Dicrophyllum marginatum)* with its white margined leaves, occurs occasionally as an escaped plant.

For the spurges of the more restricted habitats see the chapters on the dunes and the sandhills.

WATER STARWORT FAMILY *(Callitrichaceae)*

243. **Water starwort** *(Callitriche)*. Aquatic rooted plants with opposite submersed and floating leaves, the latter forming a rosette suggesting the common name: each small blade is 3-veined; the under water leaves are linear and cleft near the end, the floating leaf-blades are nearly oval: flowers inconspicuous located in the leaf axils. Three species. The water purslane *(Didiplis)* much resembles this genus but may be distinguished from it by its narrow, nearly linear floating leaves. In quiet shallow water and sometimes on wet soil. Spring and summer.

Box Family *(Buxaceae)*

244. Mountain spurge *(Pachysandra procumbens).* Perennial, 6-12 in. high, with the stamens and carpels borne in separate white flowers in the same spike; stamen stalks thickened, prominently exserted from the perianth; sepals 4, petal-like, petals absent: leaf-blades varying from ovate, margin entire or with a few blunt teeth, tapering below into a short petiole, evergreen. In shady mountain woods. Spring.

The related Japanese spurge *(P. terminalis)* is becoming widely used as a ground cover.

Mallow Family *(Malvaceae)*

245. Velvet leaf *(Abutilon Abutilon).* Annual weed, 3-6 ft. high, with solitary yellow slender-stalked flowers from the upper leaf axils; sepals united; carpels many, over 12, forming a prominent whorl in fruit, each with a sharp pointed tip: leaf-blades heart-shaped, covered with a soft velvety pubescence. In waste areas; a weed from South Asia. Late summer and fall.

246. False mallow *(Sida).* Tough weed-like plants, 1-4 ft. high, with small yellow flowers borne 1-4 from the upper leaf axils; sepals united, carpels 5: leaf-blades in our species, simple, ovate or linear, toothed: stems slender but extraordinarily strong. In waste areas, more common eastward. Summer.

247. Poppy-mallow *(Callirrhoe triangulata).* Perennial, 1-2½ ft. tall, with showy dark purple flowers 1-2 in. dia. in a few-flowered loose terminal cluster; petal blades triangular in outline, tapering to the point of attachment; stamen stalks united into a tube around the style: basal leaf-blades heart-shaped, the upper ones deeply lobed, pubescent. In grassy areas, of the western region. Summer.

St. John's-wort Family *(Hypericaceae)*

248. St. Peter's-wort *(Ascyrum).* Low to medium perennials of a shrubby nature with terminal yellow flower clusters;

sepals 4, the 2 outer much enlarged and persistent; styles 2-4, distinct: leaves small never over 1½ in. long, entire, opposite and sessile. Chiefly in moist sandy soils. Three species. Summer and fall.

249. **St. John's-wort** *(Hypericum)*. Herbaceous and shrubby plants somewhat similar to the preceding genus in aspect, but readily distinguished by the 5 petals instead of 4: stamens numerous in 3-8 groups; styles 3-5, distinct; sepals 5 nearly alike: leaves commonly with "dots" (glands) in them; simple, oval or elliptical, entire, sessile, opposite. About 20 species of which half are herbaceous. In woods and bog lands. Spring and summer.

250. **Pineweed** *(Sarothra gentianoides)*. Small, bushy branched annual, 4-20 in. high, with the tiny yellow flowers scattered on the branchlets; styles 3, distinct: branches green, leafless, the leaves reduced to scales. In sandy or dry soils, common in old fields where they appear like miniature bushes borne on a single stem at the base. Spring to fall.

251. **Marsh St. John's-wort** *(Triadenum)*. Very similar to Hypericum but differing in the pink or purplish flowers and fewer stamens (9 organized 3 groups of 3 in. each): leaf-blades large, 1-6 in. long with veins prominently bending forward along the margins; dotted with glands. Fresh water marshes. Summer.

Rock-rose Family *(Cistaceae)*

252. **Frost-weed** *(Helianthemum)*. Perennials with stems woody at the base bearing apetalous flowers at the same time or later in addition to the yellow showy petal-bearing ones; stamens many (over 12) mostly spreading laterally thus lying against the petals; stigma sessile on ovary, 3-lobed; leaves simple, blades elliptical or narrowly so, short petioled. In sandy or dry soils, chiefly eastern. Summer.

The common name is derived from the reports that during the early frosts, peculiar ice structures are produced from the base of the plant.

253. Pin-weed *(Lechea).* Bushy-branched perennials, 6-30 in. high, with woody bases bearing very small flowers scattered thickly over the branches, petals 3, sepals 5: stigmas 3, sessile (as seen under lens): leaves oblong or linear, not over 3/4 in. long, sessile. Six species in dry soils, especially common in old fields. Summer.

VIOLET FAMILY *(Violaceae)*

254. Violet *(Viola).* Well known herbs with bilateral flowers; anthers united, the stalks of the 2 lowest stamens bear basal appendages which protrude into the short spur of the lower odd petal; with the exception of the bird's foot violet (*V. pedata,* fig. 176) inconspicuous flowers without petals are formed which are highly productive of seed. These commonly appear later than the showy petal-bearing ones. The violet species range through a wide variety of habitats from dry woods and open areas to bog lands. Spring and early summer.

KEY TO THE COMMON VIOLETS

I. Petals white, with or without purplish lines.
  a. Leaves not basal.
    II. Stipules very small, not leaf like.
      Striped violet *(Viola striata).*
    II. Stipules as large as the leaves, deeply cut.
      Johnny jump-up *(Viola Rafinesquii).*
  a. Leaves basal.
    III. Leaf-blades heart-shaped.
      b. Lateral petals bearded. Bearded white violet *(Viola incognita).*
      b. Lateral petals not bearded, upper 2 petals alike.
        Sweet white violet *(V. blanda).*
    III. Leaf-blades not heart-shaped; base tapering into petiole.
      c. Leaf-blades oval. Primrose violet *(V. primulifolia).*
      c. Leaf-blades narrow, lance-shaped. Bog violet *(V. lanceolata).*
I. Petals yellow.
  d. Leaves basal: mountain species
    Round leaved yellow violet *(V. rotundifolia).*
  d. Leaves not basal.
    IV. Leaf-blades 3-lobed; mountain species.
      Tri-lobed leaf yellow violet *(V. tripartita).*

IV. Leaf-blades not lobed but heart-shaped at base.
e. Plants softly pubescent. Downy yellow violet *(V. pubescens)*.
e. Plants nearly smooth.
V. Basal lobes not flaring.
Smooth yellow violet *(V. eriocarpa)*.
V. Basal lobes laterally flaring, halberd-shape.
Halberd-leaved violet *(V. hastata)*.
I. Petals violet.
f. Small plant with large leaf-like, deeply cut stipules.
Johnny jump-up *(V. Rafinesquii)*.
f. Without large leaf-like stipules.
VI. Leaf-blades cut or lobed.
g. Leaf-blades deeply palmately divided to near base of blade; lobes narrow, finger-like. Bird's-foot violet *(V. pedata)*.
g. Leaf-blades lobed, the lobes not extending to the base.
VII. All leaf-blades lobed or blades of early leaves coarsely toothed. Early blue violet *(V. palmata)*.
VII. Blades of early and late violet leaves not lobed.
Tri-lobed leaf blue violet *(V. triloba)*.
VI. Leaf-blades not lobed or cut.
h. Leaves not basal; mountain species.
VIII. Corolla white within, violet without.
Canada violet *(V. Canadensis)*.
VIII. Corolla violet both without and within.
i. Plant finely pubescent; spur as long as the flower.
Long spurred violet *(V. rostrata)*.
i. Plant very smooth; spur short.
Dog violet *(V. conspersa)*.
h. Leaves all basal. Stemless purple violets. A group of 7 or 8 species much alike.

255. Green violet *(Cubelium concolor)*. Perennial, 1-2 ft. high, with small short-stalked greenish flowers (1/4 in. long) distributed 1 to 3 in the upper leaf axils; petals nearly alike, the lower one short-spurred: leaf-blades narrowly ovate tapering at each end, alternate. In moist mountain woods. Summer.

Passion-Flower Family *(Passifloraceae)*

256. Passion-flower *(Passiflora)*. Perennial, herb-like, climbing vines, bearing large flowers remarkable for the unusual structure made up of radiating filaments (corona) which

FIG. 176. BIRD'S-FOOT VIOLETS DO THEIR BIT TO BRIGHTEN THE SPRING TIME WOODLANDS.

is borne above the petals and at the base of the column formed by the united stamens and style: leaf-blades 3-lobed, those of the purplish-flowered species (*P. incarnata,* fig. 177) are finely toothed, while those of the yellow flowered one *(P. lutea)* are entire. In medium dry soil, climbing over other plants. Chiefly middle and eastern. Early summer.

CACTUS FAMILY *(Opuntiaceae)*

257. Prickly pear (*Opuntia,* fig. 178). Perennial leafless plants with jointed, flat, fleshy stems lying more or less prostrate, and bearing large pale yellow flowers; corolla borne above the ovary. In dry or sandy soils. Spring to fall.

MEADOW-BEAUTY FAMILY *(Melastomaceae)*

258. Meadow-beauty *(Rhexia).* Perennials with showy flowers having the corollas borne above the ovary on the vase-

FIG. 177. THE CORONA OF DELICATE RADIATING STRUCTURES IS THE MOST DISTINCTIVE FEATURE OF THE PASSION-FLOWER OR MAY-POP. IN THE SIDE VIEW (RIGHT) NOTE HOW THE STAMEN STALKS AND STYLE ARE UNITED INTO A COMMON STRUCTURE.

FIG. 178. AMONG THE MOST BEAUTIFULLY COLORED FLOWERS ARE THOSE OF THE PRICKLY-PEAR CACTUS. PHOTO BY H. L. BLOMQUIST.

like structure (hypanthium) which is free from the enclosed ovary. The 4 sepals spread from the rim of this hypanthium; ovary 4-celled: leaf-blades oval or narrower, commonly 3-veined from near the base.

Most of the meadow beauties are eastern bog plants. One species *(R. lutea)* is yellow, the remaining 6 species have purple corollas. Of these an especially noteworthy one is the tall savannah meadow beauty (*R. Alifanus,* fig. 57) with its wand-like stems bearing 1 or 2 large (1½ in. dia.) showy blossoms.

Loosestrife Family *(Lythraceae)*

259. Water purslane *(Didiplis diandra).* A small aquatic plant somewhat resembling the water-starwort; flowers without petals, very small, greenish, in the leaf axils: leaves narrow not over 3/4 in. long; opposite, the submersed ones narrower than those borne above the water: stem 4-angled, weak. In shallow water and on mud. Summer.

260. Marsh loosestrife *(Ammannia).* Annuals, 6-20 in. high, with 4-parted pink or purple flowers, the early disappearing petals small and contracted below; flowers clustered in the leaf axils: leaves narrow, opposite, not over 3 in. long. Fresh water marshes, eastern. Summer.

261. Tooth-cup *(Rotala ramosior).* Annual, 2-13 in. high, much resembling the preceding; flowers small in axils; the fruiting capsule being enclosed by the hypanthium suggests the common name: leaves slightly over 1 in. in length. Fresh water marshes, eastern. Summer.

262. Salt marsh loosestrife *(Lythrum lineare).* Perennial, 2-4 ft. high, with white or purple flowers borne singly in the leaf axils; hypanthium (flower base bearing the petals) slender: leaves linear not over 1 in. long, opposite. Margins of salt marshes along the coast. Summer.

263. Blue wax-weed *(Parsonsia petiolata).* Annual, 6-20 in. high, with short stalked bluish-purple flowers borne singly in the leaf axils; petals 6 unequal in relation to the bilateral nature of the flower: leaf-blades oval or lance-shaped and together with the stems covered with a sticky pubescence which suggests the common name. In dry or rocky uplands, central and western. Summer and fall.

EVENING PRIMROSE FAMILY *(Epilobiaceae)*

264. Marsh purslane *(Isnardia)*. Watery herbs with floating or (on mud) creeping stems; flowers small, solitary in the leaf axils; petals reddish, very small, often absent; ovary enclosed in a hypanthium with the 4 sepals spreading from the top: leaves tapering at each end, and toward the plant ending in a petiole, 1/2-1¼ in. long, opposite. Two species in fresh water marshes. Summer and fall.

265. False loosestrife *(Ludwigia)*. Annuals or perennials, 1-3 ft. high, erect with the small 4-parted flowers borne singly in the leaf axils; ovary enclosed by the hypanthium which is sometimes square as seen in section; petals in some species small and greenish but in the showier forms are clear yellow: leaves narrow, sessile, tapering at both ends, alternate. About 13 species distinguished by fruit and minor flower characters. Fresh water marshes, chiefly eastern. Summer.

266. Long stalked false loosestrife *(Ludwigiantha arcuata)*. Similar to the preceding except the flowers are stalked rather than sessile, and the leaves are opposite. Marshes. Spring and summer.

267. Primrose willow *(Jussiaea)*. Creeping or erect perennials, not over 3 ft. high, with rather showy yellow flowers borne singly in the leaf axils; stamens 8-12 in 2 rows; hypanthium slender: leaf-blades varying around lance-shaped, alternate. Four species in ponds and marshes. Summer.

268. Fire-weed *(Chamaenerion angustifolium)*. Tall graceful perennial, 2-8 ft. high, with showy terminal racemes of purple, bilateral 4-parted flowers; hypanthium slender not extending beyond the ovary; stamens 8, stalks expanded at the base: leaf-blades narrow lance-shape, taper-pointed, sessile or short-stalked, alternate. In dry places, especially in areas recently burnt over in the mountains. Summer.

269. Willow-herb *(Epilobium coloratum)*. Perennial, 1-3 ft. tall, with numerous small pink or white flowers in the upper leaf axils; corolla not over 1/4 in. wide: leaf-blades

FIG. 179. ALL OF THE FOUR-PETALED EVENING PRIMROSES ARE ATTRACTIVE FLOWERS. PHOTO BY H. L. BLOMQUIST.

narrow lance-shaped, reduced above, commonly purplish tinged. In moist soil of low areas, chiefly in the mountain region. Summer.

270. Evening primroses *(Onagra, Oenothera, Kneiffia,* fig. 179). Eleven species of closely related plants with terminal racemes of showy yellow 4-parted flowers. Leaf-blades varying around narrowly lance-shape: some are chiefly lobed; alternate.

All are erect in habit except the spreading seaside evening primrose *(Oenothera humifusa)* and the partly spreading cut-leaved species *(O. laciniata)* so familiar in old fields. The tallest one is the western or mountain region common evening primrose *(Onagra biennis)* which may reach 6 ft. The remaining 8 species belong to the genus *Kneiffia* which has 4-winged fruits together with a slender hypanthium (calyx-tube), the plants seldom being more than 3 ft. high. Mostly on dry uplands. Spring, summer, and fall.

271. Wild herb-honeysuckle *(Gaura).* Tall herbs, 2-5 ft. high, branching above, the branches ending in racemes of 4-petaled bilateral flowers; petals extending upward, the stamens relatively long and hanging downward; flowers white turning pink in age: leaf-blades narrow, remotely and inconspicuously toothed. Three species on dry uplands, central and western. Spring, summer, and fall.

272. Enchanter's nightshade *(Circaea).* Perennials, 3-24 in. high, with soft, watery stems, bearing racemes of small, white flowers; petals 2, deeply notched in the small *C. alpina*: leaf-blades broadly lance-shaped or ovate with a wavy-toothed margin, opposite. Cool mountain woods. Summer.

Water Milfoil Family *(Gunneraceae)*

273. Mermaid weed *(Proserpinaca).* Aquatic plants or reclining in shallow water with inconspicuous greenish flowers: leaves pinnately divided except the above water ones of *P. palustris,* which are narrowly elliptical; alternate. Two species. In quiet water, eastern. Summer.

274. Parrot's-feather *(Myriophyllum).* Aquatic plants with horizontal floating stems with minute flowers of two kinds, stamen bearing and carpel bearing; under water leaves pinnately finely divided mostly in whorls of 3 to 6; above water

FIG. 180. A SMART WEED IN A NEW SENSE; THE INTELLIGENCE PLANT WHICH IS SUPERSTITIOUSLY SUPPOSED TO FURNISH A BRAIN STIMULANT.

leaves small, narrow not divided, also in whorls. In quiet shallow water, central and eastern. Summer.

CARROT FAMILY *(Ammiaceae)*

275. Water pennywort *(Hydrocotyle)*. Low creeping perennials, under 8 in. high, with globular umbels of small greenish flowers: leaf-blades round in outline with scalloped margin in most species, attached to stalk at center of blades; leaves scattered along creeping stems: 2 species, including the mountain *H. Americana,* have the blades kidney-shaped and they are attached to the petiole at the base of the notch. In perennially damp grounds, especially in the coastal plain. Spring to fall.

276. "Intelligence plant" (*Centella repanda,* fig. 180). Low perennial, not over 1 ft. high, creeping, with greenish flowers in globular umbels similar to the preceding genus: leaf-blades ovate with wavy margin and squarish base, borne

FIG. 181. THE HARD BALL-LIKE HEADS OF BUTTON-SNAKE-ROOT COMPETE IN HEIGHT WITH THE TRUMPET LEAVES.

on slender clustered petioles. In moist grounds especially near the coast. Summer.

This plant is widely distributed through the tropical regions. In India there is a superstition that tea made from the leaves of this plant acts as a brain stimulant, hence the common name.

277. Snake-root *(Sanicula)*. Perennials, 1-4 ft. high, with small umbels of greenish white or purplish flowers at ends of the main bracted umbel branches; fruits distinctive, being thickly covered with hooked bristles: leaves palmately 5-parted, the outermost divisions again partly divided. Four species in upland woods. Spring and summer.

278. Button snake-root (*Eryngium*, fig. 181). Stiff, rigid perennials 1-5¼ ft. tall with greenish flowers in very compact heads; sepals persistent, projecting as short spines: leaves simple, toothed, generally appressed against the stem. Three species in open bog lands or marshes. Summer.

279. Queen Anne's-lace *(Daucus Carota).* Biennial weed 1-3 ft. high with small, white (rarely rose or yellowish) flowers in a flat-topped cluster of umbels. Central flower of the central umbel often dark purple; the two-parted fruit bristly: leaves compound, the blades much cut and divided; stem coarsely pubescent. In waste areas but not common eastward. Adventive from Europe. Summer.

280. Sweet cicely *(Washingtonia).* Perennials, 1-3 ft. high, with compound umbels bearing few white flowers: leaves compound, the primary divisions being divided again into leaflet bearing stalks: root with odor of anise which is stronger in the smooth-leaved *W. longistylis.* One other species *(W. Claytonii)* having pubescent foliage found in rich woods, chiefly in the mountains. Spring.

281. Harbinger-of-spring *(Erigenia bulbosa).* Small perennial, 3-6 in. high, with the 1-4 primary umbel rays bearing a small few flowered umbel of white flowers with dark anthers: leaves twice divided, the primary divisions 3 in number: underground stem a tuber. In mountain woods. Early spring.

282. Scaly seed (*Spermolepis divaricatus,* fig. 182). Slender stemmed branching annual, under 2 ft. high, with small umbels of white flowers borne wide apart: leaves 2 or 3 times divided, the leaflet blades very narrow or thread-like. In sandy soils, especially in the coastal plain. Spring.

283. Chervil *(Chaerophyllum).* Annuals, 6-30 in. tall, with small, white flowers in few-rayed umbels; fruit about 2½ times as long as broad, nearly smooth: leaves compound alternate, the leaflets much cut and divided, pubescent. Two species native in upland soils often in waste places. Spring and summer.

284. Alexanders *(Zizia).* Perennials, 6 in.-3 ft. high, with yellow flowers in twice compound umbels; no whorl of bracts at base of inflorescence; basal leaves simple, leaf-blades ovate, heart-shaped in *Z cordata;* in the other 2 species the leaves are of 3 or sometimes 5 leaflets all rather finely toothed. Woods and meadows generally but especially in the piedmont and mountains. Spring and summer.

285. Water hemlock *(Cicuta).* Tall perennials, 3-6 ft. high, with large twice compound umbels of white flowers: leaf-blades large 2-3 times pinnate on long sheathing petioles: stems generally striped with purple; base of stem with cavities in pith, as seen in vertical section. Two species, found only in marshes. Summer.

286. Honewort *(Deringa Canadensis).* Perennial, 1-2½ ft. high, with spreading branches bearing few flowered compound umbels of small white flowers: leaves of 3 leaflets, each between 1½ and 4 in. long, smooth with irregularly toothed margin. In mountain woods. Spring and summer.

287. Yellow pimpernel *(Taenidia integerrima).* Perennial, 1-3 ft. high, with compound umbels of yellow flowers: leaves pinnate, 2-3 times divided, the divisions chiefly in threes; leaflets ovate, with entire margins. In dry soil, chiefly western. Spring.

288. Water parsnip (*Sium cicutaefolium*, fig. 182). Tall perennial, 2-6 ft. high, with a whorl of small bracts at the base of the compound umbel of white flowers: leaves much divided, the leaflets narrow (linear) and toothed; under-water leaves rather finely divided; leaf stalks prominently sheathing the stem. Marshes. Summer.

289. Bishop-weed *(Ptilimnium capillaceum).* Annual, 1-2 ft. high, with slender stems bearing delicate compound umbels of white flowers: leaves much divided into thread-form leaflets. In wet soil and brackish marshes, near the coast. Summer.

290. Joint-leaf *(Lilaeopsis lineata).* Low creeping plants, 2-9 in. high, with the white-flowered umbels borne on slender stalks from the spreading stems: leaves arising in twos or threes from the stem joints, each leaf a slender club-shaped

FIG. 182. THE WATER PARSNIP (LEFT) AND SCALY SEED (RIGHT) FLOWERS SHOW THE CHARACTERISTIC UMBEL ARRANGEMENT OF THE CARROT FAMILY.

jointed structure with partitions at the joints. Two species in fresh water or salt marshes. Summer.

291. Nondo *(Ligusticum Canadense)*. Tall and branched perennial, 2-6 ft. high, the branches terminating in large compound umbels of white flowers; 2-6 narrow bracts at base of each inflorescence: lower leaves large (1 ft. wide) the primary divisions in threes; leaflets ovate, coarsely toothed. In mountain woods. Summer.

292. Meadow parsnip *(Thaspium)*. Perennials, 1-4 ft. high, with compound umbels of yellow flowers (purple in *T. trifoliatum*): leaves various; mostly 3 leaflets in the purple meadow parsnip, in the other 2 species the leaves are twice compound with the ultimate leaflets in threes. Upland woods and river flood plains, chiefly in the piedmont and mountains. Summer.

293. Angelica *(Angelica)*. Tall, slender perennials, 2-6 ft. high, with twice compound umbels of white flowers terminating the branches; no bracts at base of the inflorescence: leaves large, compound with toothed leaflets. Two species: *A. villosa* is pubescent above and *A. Curtisii* a strictly mountain species which is smooth or nearly so. In woods. Summer.

294. Cow-bane *(Oxypolis).* Tall, very slender perennials, 2-6 ft. high, with few-flowered ultimate umbels of white flowers: leaves of the bog species *O. filiformis* simple, being represented only by sheathing, taper-pointed petioles; leaves of the other 3 species including the marsh *O. rigidior,* with 3 or more narrow leaflets. Marshes, bogs and wet land, eastern. Summer.

295. Cow parsnip *(Heracleum lanatum).* Very tall, coarse perennial, 4-8 ft. high, with correspondingly large compound umbels of white flowers: primary leaf divisions in threes, the leaflet blades strongly pubescent beneath; stem also hairy; fruit notched at end. In wet ground, chiefly western. Spring and summer.

Wintergreen Family *(Pyrolaceae)*

296. False wintergreen *(Pyrola).* Two species of low perennials with flowering stems not over 20 in.; petals white or greenish white, wax-like; raceme borne flowers on short stalks, frequently nodding; stamens 10: leaves all basal, blades broadly oval, margin finely toothed if at all. Chiefly in mountain woods. Summer.

297. Spotted wintergreen (*Chimaphila,* fig. 183). Perennials, not over 1 ft. high, with terminal inflorescence of a few white or delicate pink flowers with wax-like petals: leaves of *C. maculata* lance-shaped, variegated with pale green along the veins; *C. umbellata* has leaves oblance-shaped (broader above the middle) and not variegated; blades of both are thick. Two species in upland woods chiefly of the piedmont and mountains. Summer.

FIG. 183. AN INDIAN NAME "PIPSISSEWA" HAS CLUNG TO THE SPOTTED WINTERGREEN; BOTH NAMES ARE IN COMMON USE. PHOTO BY H. L. BLOMQUIST.

INDIAN PIPE FAMILY *(Monotropaceae)*

298. Indian pipe (*Monotropa uniflora,* fig. 184). Remarkable, erect, low herbs without green or other color being present at first, so that the white cellulose of the cell walls of the stems, reduced leaves and the single nodding flower, is all that is seen. In age they first turn pink then as the cells die, the plant rapidly turns black. The ovary is composed of 5 united carpels; the petals are 3-6 in number and are united below. Unable to make its own food like the green plants, it obtains its nutrition from the organic matter of the rich forest soil or from the roots of living plants, as is held by some. In moist, deep woods chiefly of the piedmont and mountain regions. Summer.

299. Pine-sap *(Hypopitys).* Root parasites, not over 14 in. high, with a terminal raceme of a few flowers; petals distinct, each with a small swelling at the base: leaves reduced to scales. *H. Americana* is lemon-yellow or pinkish and very finely pubescent while *H. lanuginosa* is reddish brown and

FIG. 184. THE GHOST OF THE WILD FLOWER COMMUNITY. INDIAN PIPES ON TOP OF CLINGMAN'S DOME IN THE SMOKY MOUNTAIN PARK. NOTE TWIGS OF BALSAM IN FOREGROUND.

strongly pubescent. In deep woods, especially in the mountains. Summer.

300. Carolina beech drops *(Monotropsis odorata).* Very small root parasite, 2-4 in. high, with a few small flowers borne terminally; petals united; stamens 10; leaves reduced to scales. The plant is light purplish brown and the flower bearing stalks arise from the ground in a cluster. Only found in deep woods near the summits of our higher mountains. Very rare. Early spring.

HEATH FAMILY *(Ericaceae)*

Though all the members of this family are shrubby in nature, two low species are included here because of their interest and value as wild flowers.

301. Trailing arbutus (*Epigaea repens,* fig. 185). Prostrate or creeping woody plants with terminal and lateral clus-

FIG. 185. THE TRAILING ARBUTUS BREATHES THE MOST FRAGRANT BREATH OF SPRING. PHOTO BY A. F. ROLLER.

ters of white or pink very fragrant flowers; corolla tube hairy within; ovary 5-parted: leaf-blades elliptical, short-stalked, evergreen. In upland woods of widely divergent character being found in our sunny sandhills, as well as the shady broad-leaved forests and conifer woods of the mountain slopes. Early spring.

302. True wintergreen *(Gaultheria procumbens)*. Small, evergreen perennial, not over 6 in. high, with upright branches from woody trailing stems, each branch bearing a few bell-shaped flowers from the leaf axils; petals recurved from edge of corolla tube: leaves forming a rosette at end of the erect branch, leaf-blades ovate, shining, with the flavor of wintergreen. In upland woods and occasionally in shrub bogs of the eastern region. Spring and summer.

Blueberry Family *(Vacciniaceae)*

303. Trailing blueberry *(Vaccinium crassifolium).* Perennial, creeping plant with small clusters of red-tinged flowers; corolla bell-shaped: leaf-blades varying from oval to orbicular, evergreen, alternate: stems woody and vine-like, spreading over the ground. In sandy bog or moist lands, eastern. Spring.

Pyxie Family *(Diapensiaceae)*

304. Pyxie *(Pyxidanthera).* Very low, creeping perennials with white flowers solitary and terminating the numerous branchlets; stamens 5, attached to the corolla tube; pistil of 3 united carpels: leaves small, evergreen more or less overlapping. Two species: *P. brevifolia* (Fig. 83) forms flat compact mats in deep, dry sand; leaves very short, seldom over 1/16 in. long. Rare species only known from vicinity of Spout Springs, Harnett County, North Carolina. *P. barbulata* occurs in moist sand areas of the lower coastal plain and has longer leaves and larger flowers. Early spring.

Galax Family *(Galacaceae)*

305. Shortia *(Shortia galacifolia).* Low evergreen perennial with flowering stalks from ground, not over 6 in. high; flowers solitary; petals 5, white or pink, marked with lines; 5 scale-like structures at base of corolla tube in addition to the 5 stamens: leaf-blades broadly oval to orbicular with a finely wavy-toothed margin, thick, evergreen. In the vicinity of mountain streams. Very rare. Spring.

306. Galax (*Galax,* fig. 186). Low evergreen perennial with flower stalks and leaves arising from scaly rootstalks; flowers white, many in a slender spike-like raceme 8-14 in. high; tube of corolla short, bearing the stamens: leaves orbicular, fine wavy-toothed, leathery becoming bronze-color in the fall; in size the leaves vary widely from 1-4 in. dia. On mountain sides and cool steep north-facing slopes in the piedmont and coastal plain. Spring and summer.

FIG. 186. THE GALAX SPIKES SHOW UP LIKE CANDLES ON A BIRTH-DAY CAKE. PHOTO BY A. F. ROLLER.

PLUMBAGO FAMILY *(Armeriaceae)*

307. Sea lavender *(Limonium).* Perennial, 6-24 in. high, with small flowers in terminal much branched inflorescence; sepals 5, united into a 10-ribbed tube; petals 5, contracted below with stalks extending to base of calyx tube: leaf-blades fleshy, oval, tapering into the long petioles. Two species, *L. Brasiliense* has white flowers and *L. Carolinianum* has pale purple ones. In salt marshes. Summer.

PRIMROSE FAMILY *(Primulaceae)*

308. Feather foil *(Hottonia inflata).* Curious aquatic herb with inflated inflorescence-bearing stalks bearing the small, white flowers in whorls at the nodes: in contrast to the swollen flower stalks which are borne above the water, the finely divided leaves are all submersed and arranged in whorls at the base of the cluster of flower stalks. In quiet water, eastern. Summer.

309. Water pimpernel *(Samolus floribundus).* Perennial, 5-24 in. high, pale green, slender stems bearing tiny, white flowers (1/8 in. dia.) in racemes: leaves alternate, the blades oval, tapering into winged petioles. In marshes and wet soil especially common near the coast. Spring and summer.

310. Loosestrife *(Lysimachia).* Erect perennials, 1-3 ft. high, with a leafy terminal inflorescence; petals 5, yellow, spreading from the top of a short corolla tube; stamens 5, with no reduced stamens (staminodia) alternating with them: leaves opposite or whorled, the blades broadly lance-shape to narrower, entire. Five species of which *L. quadrifolia* and *L. asperulaefolia* with their sets or whorls of 3 or 4 leaves at a node are very distinctive. The latter species is found only in the pine areas of the coastal plain and has the flowers borne mostly in the upper part of the plant, while the former species is widely distributed in woods of the piedmont and mountains and has the flowers borne in the leaf axils throughout most of the plant. *L. terrestris* with opposite leaves is a marsh plant. Summer.

311. Sterile stamen loosestrife *(Steironema).* Erect perennials similar to the preceding but bearing flowers having 5 reduced stamens without anthers, alternating with the 5 normal ones: leaves variable in our 6 species. In woods or marshes *(S. radicans).* Summer.

312. American cowslip *(Dodecatheon).* Very attractive perennials with white or light purplish flowers in an umbel from the single stalk arising from the short rootstock; flowers nodding 1/2-3/4 in. long, the 5 petals strongly reflexed, the stamens exserted: leaf-blades oval, tapering into the petiole below; clustered in a rosette. Two species, on wet cliffs chiefly in the mountains. Spring.

Logania Family *(Spigeliaceae)*

313. Mitrewort *(Cynoctonum).* Erect branched perennials not over 28 in. high, with 4-angled stems terminating above in short spikes of small, white flowers; corolla tube very short (not over 1/8 in.) constricted above, petals striped: leaves opposite, blades ovate to narrowly lance-shaped. Two species in wet soil of bogs or marsh borders, central and eastern. Summer.

314. Carolina pink *(Spigelia Marylandica).* Attractive erect perennial, 1-2 ft. high, with terminal 1-sided spikes of scarlet flowers; corolla tube slender (1½-2 in. long) with the 5 narrow lobes radiating from the summit, tube yellow within: leaves sessile, opposite, the blades oval or lance-shape. Mountain woods. Summer.

315. Spreading wire-plant *(Polypremum procumbens).* Spreading much branched annual weed-like herb with inconspicuous white flowers in terminal bracted clusters; flowers mostly 4-parted; corolla bearded in the short throat: leaves linear, opposite: stems slender, rigid, 4-angled. In sandy soil, chiefly in the coastal plain. Summer.

Gentian Family *(Gentianaceae)*

316. Sabbatia *(Sabbatia).* Erect branching perennials, 5-30 in. high, with very attractive rose-pink or white flowers borne solitary or in an open cluster on the upper part of the plant; petal number widely variable in the genus (4-12); petals united below, the tubular part of the corolla being differently colored, giving the flower a central "eye"; stigmas 2, but only one cell (cavity) in the ovary: leaves opposite, the blades varying around lance-shape. About 9 species chiefly in moist soil and bog lands, mostly in the coastal plain. Summer.

KEY TO SPECIES

Flowers white: eastern wet lands species.
  All branches opposite on the stem.

FIG. 187. THE WHITE SABBATIA IS AN ESPECIALLY ATTRACTIVE WILD FLOWER.

FIG. 188. THE MOUNTAIN SABBATIA BRIGHTENS ITS CORNER OF THE WOODS WITH ITS DELICATE ROSE TINTS.

Two species much alike. *S. lanceolata and paniculata*, fig. 187.
All or nearly all branches alternate.
Sepals much shorter than the petals. *S. Elliottii.*
Sepals as long as the petals. *S. stellaris.*
Flowers rose or pink: eastern and western species.
Petals 8-12. *S. dodecandra*, fig. 58.
Petals 4-5.
Branches all opposite: chiefly western species.
Larger leaves with narrow blades. *S. angustifolia.*
Larger leaves with blades ovate or broader.
S. angularis, fig. 188.
Branches chiefly alternate: eastern species.
Sepals narrow, leaf-like, longer than the corolla lobes: coastal species. *S. calycina.*
Sepals narrow but not prominent; not longer than the corolla lobes.
Style shorter than the stigmas. *S. stellaris.*
Style longer than the stigmas. *S. campanulata.*

317. Ague-weed *(Gentianella quinquefolia)*. Perennial, 1/2-3 ft. high, with stiff angular or 4-winged stems bearing compact terminal and axillary clusters of the narrow, tubular, purplish-blue flowers; sepals very narrow, about 1/3 as long as the 3/4 in. long corolla tube; petals merely 5 teeth on rim of the corolla tube: leaf-blades ovate or lance-shape, leaves not much over 3 in. long, opposite. In moist woods, ranging up to 6000 ft. in the mountains. Summer and fall.

318. Smooth gentians *(Dasystephana)*. Erect low to medium perennials, with compactly clustered tubular, blue or purple flowers, the corollas showing extra minutely toothed folds or plaits between the lobes or petals proper: leaves opposite, the blades ovate, lance-shaped or linear, entire. Of especial interest are: the sandhill gentian (*D. Porphyrio*, fig. 189) of moist sandy soils in the coastal plain. It is to be recognized by its very narrow (linear) leaves and sky-blue, widely flaring corolla; the closed gentian *(D. Andrewsii)* with very minute corolla lobes has the plaits and the corolla tube contracted at the end, leaving a small opening. Eight

FIG. 189. IN THE BLUE OF THE SANDHILL GENTIAN, WE HAVE A MIRROR OF THE SKY.

species in damp, shady places or moist soil in the open. Summer and fall.

319. Fringed gentian *(Anthopogon crinitum)*. A very attractive species, 1-3 ft. high, with solitary blue flowers terminating the branch ends; flowers mostly 4-parted the margins of the spreading 4 petals being prominently fringed; sepals 4, the two outer larger than the 2 inner: leaves opposite, the blades lance-shaped, tapering above. In deep mountain woods, and moist meadows. Late summer and fall.

320. American columbo *(Frasera Carolinensis)*. Tall perennial, 3-7 ft. high, with a large terminal loose cluster (1-2 ft. long) of yellowish white spotted, 4-parted flowers; the corolla lobe or petal bears a curious fringed gland just below the middle: leaf-blades lance-shaped, the basal ones tapering into slender petioles. In dry woods. Summer.

FIG. 190. THE MODEST FLOWERS OF THE PENNYWORT PEEP OUT AMID THE SOMBER SHADOWS OF DEEP WOODS.

321. Bartonia *(Bartonia)*. Low and rather insignificant annuals or biennials, 6-16 in. high, with slender leafless stems bearing small, yellow or white flowers in a raceme or a branching inflorescence above; stamens attached just below the notches between the corolla lobes: leaves changed to scales which lie appressed against the stem. The pure white flowered form is *B. verna* and the yellowish-white ones belong either to the unbranched *B. Virginica* or the much branched *B. lanceolata.* In moist soil or open bogs chiefly in the coastal plain. Spring and summer.

322. Pennywort (*Obolaria Virginica,* fig. 190). Small perennial, 2-8 in. high, compact herb with the pink or white flowers sessile in the leaf axils, the lowest leaves being reduced to scales; flowers with 2 sepals and a 4-lobed corolla tube with the small stamens attached near the notches: leaves somewhat wedge-shape, commonly tinged with purple. In deep woods

FIG. 191. THIS GREEN-FLOWERED RELATIVE OF THE SHOWY BUTTERFLY WEED IS FREQUENT ON SAVANNAHS.

partially hidden among the fallen leaves, chiefly in central portion of the state. Spring.

FLOATING HEART FAMILY *(Menyanthaceae)*

323. Floating heart *(Limnanthemum)*. Water plants with small floating leaves resembling those of the water-lily, round with a deep notch; the white flowers each have a 5-lobed corolla with 5 stamens attached to it. Two species: *L. lacunosum* has flowers not over 1/2 in. broad while *L. aquaticum* has larger flowers mostly 3/4 in. broad. In quiet, open water, especially in the coastal plain. Spring to fall.

DOGBANE FAMILY *(Apocynaceae)*

324. Twin-fruit *(Amsonia)*. Perennials, 1-4 ft. high, commonly with a number of branches from the root-stock, each bearing a showy loose cluster of light purplish blue flowers above; corolla tube narrow with the 5 petals widely flaring at the top; carpels 2 developing in fruit into two divergent slender pods: leaves alternate, the blades of the middle and western woodland species *(A. Amsonia)* are narrowly lance-shape while those of the other smaller species *(A. ciliata)* are slender, linear, almost thread-form. The latter plant is only found in the coastal plain, especially in the sandhills. Spring and summer.

325. Dog-bane *(Apocynum).* Much branched perennials, 1-4 ft. high, with the branches ending in few flowered clusters of small, white or pink blossoms; calyx short, 5-lobed, corolla bell-shaped with 5 erect or recurving petals; fruit, 2 separate pods: leaf-blades ovate or lance-shape. Two species; the one with pink flowers and recurved petals is *A. androsaemifolium;* the other with greenish-white corollas and erect petals is *A. cannabinum.* In open woods and fields. Summer.

MILKWEED FAMILY *(Asclepiadaceae)*

326. Fragrant yellow milkweed *(Podostigma pedicellata).* Flowers of the milkweed type (see milkweed below) but with erect petals; in umbels, corolla 5-angled at base, supporting a crown (accessory structure) of 5 incurved pitcher-like bodies with inrolled margins: leaf-blades narrow, lance-shape, practically sessile, opposite. In coarse sandy soil, eastern. Not common. Summer.

327. Green milkweed (*Acerates,* fig. 191). Flowers of the milkweed type (see milkweeds *(Asclepias)* below) but with no horn inserted in the 5 curious inrolled petal-like structures which stand erect above the series of purplish-green true petals which are strongly reflexed or turn downward in maturity hiding the sepals: leaf-blades lance-shaped *(A. viridiflora)* to linear *(A. floridana),* leaves both alternate and opposite on the same plant. Moist soil and savannahs. Summer.

328. Milkweed *(Asclepias).* Flowers in terminal or lateral umbels; corolla with spreading or reflexed lobes (petals);

FIG. 192. THE BUTTERFLY WEED IS ONE OF THE SHOWIEST MEMBERS OF THE MILKWEED FAMILY. PHOTO BY H. L. BLOMQUIST.

in addition to the corolla these flowers have a superstructure called a corona consisting of 5 curved funnel-like structures in the center of each of which is an incurved "horn," a tiny pointed structure. These 5 units of the corona called "hoods" are united below into a column; the carpels and stamens are united with a common structure in the center. By inserting a pin to imitate an insect leg, the pollen may be removed from the slits in the upper part in paired pollen masses: leaf-blades varying from broadly elliptical to linear. About 14 species of which the most striking are keyed out below, the color being based on the corolla lobes. In a wide range of soils extending into marshes. Summer.

KEY TO THE MILKWEEDS

Flowers white.

All leaves in whorls of 3-6. Whorled milkweed *(A. verticillata)*.

Middle leaves only in whorls of 4.

Four leaved milkweed *(A. quadrifolia)*.

No leaves whorled. White milkweed *(A. variegata)*.

Flowers orange.

Stems spreading, leaf-blades elliptical.

Spreading butterfly weed *(A. decumbens)*.

Stems erect; leaf-blades narrowly lance-shape.

Common butterfly weed (*A. tuberosa*, fig. 192.)

Flowers yellow; a variety of the preceding common in the sandhill region.

Flowers greenish.

Stems with definite fine close pubescence.

False green milkweed *(A. aceratoides)*.

Stems smooth-except possibly too faint lines of pubescence.

Poke milkweed *(A. exaltata)*.

Flowers red or deep reddish purple. Marsh milkweed *(A. rubra)*.

Flowers purple or pink.

Plant spreading appearing as if reclining on its side, resting on the coarse heavy mottled leaves. Sandhill milkweed *(A. humistrata)*.

Plants erect. About 6 species of upland woods and meadows distinguished by minor differences such as leaf shape.

329. Sand vine *(Gonolobus laevis)*. A twining herbaceous vine with axillary umbels of the milkweed type of flower (see milkweeds *(Asclepias)* preceding); the white petals erect longer than the hoods: leaves opposite, the blades heart-shaped, delicate, 3-7 in. long. Along rivers and moist upland. Summer.

330. Salt marsh milkweed (*Seutera palustris*, fig. 193). Twining vine with axillary umbels of greenish white or purplish milkweed-like flowers: distinctive among the genera of the milkweed family for its hanging linear leaves. In salt marshes. Summer and fall.

FIG. 193. NO ONE WOULD THINK THIS PLANT, THE SALT-MARSH MILKWEED *(Seutera)*, WAS A MEMBER OF THE MILKWEED FAMILY. ITS DROOPING SABER-LIKE LEAVES ARE VERY DISTINCTIVE.

DICHONDRA FAMILY *(Dichondraceae)*

331. Creeping twin fruit *(Dichondra Carolinensis)*. Low perennial, 3-5 in. high, with creeping stems bearing 1-3 small, greenish white flowers on delicate stalks from the leaf bases; flower about 1/4 in. broad; calyx covered with silky pubescence: leaf-blades kidney-shaped on slender petioles longer than the flower stalks. In moist ground and in lawns, especially in the coastal plain. Spring to fall.

MORNING GLORY FAMILY *(Convolvulaceae)*

332. Small-flowered morning glory *(Breweria)*. Trailing and twining plants with white or in one species *(B. aquatica)* purple flowers borne commonly in a cluster of 2-5 on axillary stalks; styles 2, distinct; corolla more or less plaited: leaf-blades various, elliptical to linear. Our 3 white flowered spe-

FIG. 194. THE LEAVES OF THE SALT-MARSH MORNING GLORY ARE ARROWS INSTEAD OF HEARTS. THE FLOWER IS LIGHT PURPLE.

cies are found in dry sandy soils, especially those of the sandhills, the purple flowered one in moist sandy areas. Summer.

333. Red morning glory *(Quamoclit coccinea)*. Climbing and twining with scarlet-red flowers borne singly or in clusters in the leaf axils; styles united; corolla tube narrow, 3/4-1½ in. long, spreading horn-like at the end: leaf-blades heart-shaped, with tapering tip. In river bottoms and moist soils, local. Naturalized from tropical America. Summer.

334. Morning glory *(Ipomoea and Pharbitis)*. Two closely related genera with long trailing, climbing or twining vines bearing mostly solitary, axillary blue, purple or white flowers on long stalks; flowers commonly large, showy, funnel-like with the petals indicated by the 5 short pointed tips; stigma nearly spherical: leaf-blades varying in the different species from broadly heart-shaped to narrow arrow-like deeply 3-lobed. In dry uplands and in sandy soils. Summer.

KEY TO SPECIES

Stems pubescent.
  Corolla purple. Purple morning glory *(P. purpurea)*.
  Corolla blue without and white within (sometimes all white).
    Ivy-leaved morning glory *(P. hederacea)*.
Stems smooth.
  Leaf-blades heart-shaped, flower white (sometimes pinkish).
    Wild potato vine *(I. pandurata)*.
  Leaf-blades arrow-shaped (lobes narrow), flower purple.
    Arrow-leaved morning glory (*I. speciosa*, fig. 194).

335. Bindweed *(Convolvulus)*. Trailing or climbing vines with morning glory-like flowers borne singly from the leaf axils; stigmas 2, flattened; leaves various. Chiefly in dry soils. Spring and summer.

KEY TO SPECIES

Flowers purple, stems smooth. Hedge bindweed *(C. sepium)*.
Flowers white (sometimes pinkish), stems pubescent.
  Leaf-blades slightly arrow-shaped, the basal lobes short and blunt.
    Trailing bindweed *(C. repens)*.
  Leaf-blades elliptical. Upright bindweed *(C. spithamaeus)*.

Dodder Family *(Cuscutaceae)*

336. Love vine *(Cuscuta)*. Remarkable parasitic plants of a yellow color, thickly twining over other plants to which they attach themselves by short sucker-like organs; flowers small in clusters: leaves reduced to mere scales. Some 7 or 8 species occur in our area growing on herbs and shrubs. Summer.

Water-leaf Family *(Hydroleaceae)*

337. Water-leaf *(Hydrophyllum)*. Perennials, 8-30 in. tall, with terminal loose clusters of somewhat showy blue-purple or white flowers; corolla lobes twisted in the bud; stamens longer than the corolla: leaves simple, the blades lobed and coarsely toothed or pinnate with the leaflets coarsely toothed. Four species, in moist woods, chiefly in the mountains. Spring.

FIG. 195. THE MOSS PINK IS DISAPPEARING FROM OUR ROADSIDES, NOW THAT ROCK-GARDENING HAS BECOME SO POPULAR.

338. Creeping water-leaf *(Nemophila microcalyx).* Low spreading, annual, delicate herbs with very small (1/8 in. wide) blue or white flowers borne in the leaf axils on slender stalks: leaf-blades pinnately divided into 3 or 5 lobes with coarse teeth. In moist woods, chiefly in the mountains. Spring.

339. Large calyx water-leaf *(Macrocalyx Nyctelea).* Annual, 4-12 in. high, with blue or white flowers few in racemes; calyx 5-lobed as long or slightly longer than the corolla; stamens shorter than the corolla: leaf-blades pinnately divided, the lobes coarsely toothed. In moist shady woods, chiefly in western half of the state. Spring and summer.

340. Small leaved water-leaf *(Phacelia).* Annuals or biennials, 3-20 in. high, with purple, blue or white flowers in terminal loose clusters: corolla lobes overlapping in bud, not twisted: leaf-blades variously pinnately cut, generally smaller than those of the coarser true water-leaf *(Hydrophyllum)* plants. About 6 species, mostly in moist shaded areas, chiefly in the mountains. Spring.

341. Marsh water-leaf *(Nama quadrivalvis).* Perennial, 1-2 ft. high, erect or reclining with axillary light blue flowers accompanied by slender spines; corolla tube short, the 5 lobes spreading; ovary 2-celled, the 2 styles above distinct to the base: leaf-blades lance-shaped, more or less hairy. In wet soil of marsh borders, eastern. Summer.

Phlox Family *(Polemoniaceae)*

342. Phlox *(Phlox).* Most attractive perennials, 6-48 in. tall, bearing showy terminal clusters of bright hued blue, purple or white flowers; corolla tube slender expanding sharply above with the 5 spreading petals; stamens 5 hidden in the corolla tube; calyx tube narrow, 5-ribbed: leaf-blades mostly lance-shape, reduced in certain species as the moss pinks to numerous stiff, sharp appendages. Dry and moist uplands, never in bogs. Spring and summer.

KEY TO COMMON SPECIES

Leaves reduced, short rigid, clustered.
- Calyx about ¼ in. long; plants spreading, often forming a dense ground cover.
  - Petals deeply notched. Mountain moss pink *(P. Brittonii).*
  - Petals shallowly notched. Common moss pink (*P. subulata,* fig. 195).
- Calyx about ⅜ in. long; stems clustered, erect, not spreading and forming mats. Sandhill moss pink *(P. Hentzii).*

Leaves not strikingly reduced into rigid appendages.
- Erect flowering stems, producing lateral creeping shoots from the base.
  - Leaves of lateral branches with blades broader above the middle. Creeping phlox *(P. reptans).*
  - Leaves of branches with blades broadest at middle or below. Wild blue phlox *(P. divaricata).*
- Erect flowering stems not producing creeping branches.
  - Wild phloxes. A group of 6 or 7 species ovate or lance-shaped leaf-blades. Of these *P. paniculata* is especially noteworthy as the tallest of our phloxes, often reaching 4 ft. in height.

343. Greek valerian *(Polemonium reptans).* Perennial, 1½-2½ ft. high, with showy terminal loose cluster of funnelform purplish-blue flowers; calyx teeth not tipped by bristle-point; stamens longer than the corolla: leaves pinnate, alternate, the leaflets, ovate or lance-shaped. In moist soil, especially in the western region. Early summer.

GROUND CHERRY FAMILY *(Solanaceae)*

344. Ground cherry *(Physalis).* Unattractive annuals or perennials, 1-3 ft. high, bearing in the upper leaf axils single, solitary, nodding dull yellow flowers; calyx in fruit enlarging with an angular bladder-like sack enclosing the berry, corolla tube commonly with a dark center: leaf-blades ovate or lance-shaped in outline, coarsely toothed in many of our forms. About 9 species. Uplands in both moist and dry soils. Summer.

345. Nightshade and horse-nettle *(Solanum).* Unattractive weeds with flowers in few-flowered clusters; sepals united at base, spreading; corolla tube short, the 5 petals spreading; anthers extending from corolla tube in the form of a cone: leaves various: fruit a berry. In waste places, widely distributed. Summer.

KEY TO SPECIES

Leaf-blades ovate, entire, stems not prickly, berry black.
Night-shade *(S. nigra).*

Leaf-blades coarsely toothed or lobed, stems prickly, berry yellow-orange.
Horse-nettle *(S. Carolinense).*

346. Jimson weed *(Datura).* Coarse annual weeds with large solitary short stalked white or purplish morning-glory-like flowers from the upper leaf axils; calyx base persistent, under the prickly fruiting capsule: leaf-blades coarsely toothed or lobed in *D. Stramonium* or entire in *D. Metel.* Both species introduced in waste areas; from the tropics.

BORAGE FAMILY *(Borraginaceae)*

347. Wild comfrey *(Cynoglossum Virginicum).* Coarse biennial or perennial, 1½-3½ ft. high, with a terminal loose cluster of purple or white flowers; corolla about 3/8 in. broad, the tube short and closed above by 4 scale-like structures; stamens included: leaves mostly at or near the base, the blades rough hairy, oval, the lower tapering into petioles, the upper sessile. Two species in open woodlands, western. Spring.

348. Stickseed *(Lappula Virginiana).* Rather coarse annual or biennial 2-4 ft. high, with the much branched stem ending in racemes of very small (1/8 in. wide) flowers the lower ones going into fruit; corolla funnel-form with 5 scales in the throat; fruit covered with barbed prickles, hence the common name: leaves alternate, the blades oval, the lower or basal ones disappearing at flowering time. In dry woodlands and thickets, western. Summer.

349. Bluebells *(Mertensia Virginica).* Attractive perennial, 8-24 in. high, with a loose terminal cluster of blue-purple showy flowers; corolla tube slender below, expanded above into the 5-lobed and plaited portion which is about 1/2 in. broad; stamens not longer than the corolla; anther stalks delicate: leaves alternate with blades varying around oval, entire, the lower tapering into winged stalks. On moist wooded hill and mountain slopes chiefly in the western region. Spring.

350. Wild forget-me-not *(Myosotis).* Low plants, 4-20 in., with more or less curving racemes of small funnel-form blue or white flowers, the lower ones going early into fruit; corolla tube short, stamens included; flower stalks without leaf-like structures (bracts) at base: leaf-blades narrow, rounded at tip, tapering toward sessile base, pubescent. Five species, much alike. Mostly in dry upland but 2 species prefer moist soil, central and western. Spring and summer.

351. Gromwell *(Lithospermum).* Erect annuals, biennials or perennials, 4-20 in. high, the branches ending in bracted racemes; i. e., each flower is borne in the axil of a reduced leaf (bract); corolla funnel-like with spreading petal lobes; stamens 5, included, stalks short; fruit of 4 smooth nutlets: leaves alternate, the blades lance-shape to linear, pubescent, entire. On dry uplands. Spring and summer.

KEY TO SPECIES

Flowers showy, yellow orange.
  Corolla tube bearded within at base; corolla about ⅜ in. wide.
    Showy gromwell *(L. Gmelini)*.
  Corolla tube not bearded within; corolla about ½ in. wide.
    Small showy gromwell *(L. canescens)*.
Flowers white or pinkish inconspicuous, only ¼ in. broad.
    Corn gromwell *(L. arvense)*.

352. False gromwell *(Onosmodium)*. Unattractive perennials, 1¼-4 ft. high, with curved racemes of small, dull yellowish white flowers; corolla tube rather narrow, the lobes not spreading; fruit 4 smooth and shining white nutlets, often persisting on the old dead plants: leaves alternate, the blades lance-shaped to linear, rough being covered with short stiff hairs. In dry and especially sandy soil, central including the "Sandhills." Summer.

HELIOTROPE FAMILY *(Heliotropiaceae)*

353. Heliotrope *(Heliotropium)*. Annuals, 5-28 in. tall, with the white or blue flowers in terminal curved spikes, the spikes often in pairs; petals united into a cylindrical tube, the lobes spreading above; stamens 5, included, ovary not deeply 4-lobed: leaf-blades lance-shaped or narrower in the seaside heliotrope *(H. Curassavicum)*, broader in the weed species *(H. Europaeum* and *H. Indicum)*.

VERBENA FAMILY *(Verbenaceae)*

354. Vervain *(Verbena)*. Medium to tall perennials, 1-5 ft. high, with varicolored flowers in terminal slender spikes; corolla funnel-form with 5 spreading lobes; stamens 4, 1 pair longer than the other pair; fruit consists of 4 elongated nut-

FIG. 196. THIS LITTLE VERBENA-LIKE FLOWER CALLED FOG-FRUIT MAY BE FOUND READILY IN THE SANDS ALONG THE SALT-MARSH MARGIN.

lets: leaves opposite, the blades lance-shaped or narrower, coarsely toothed or lobed and more or less hairy. Five species which hybridize rather freely. On upland dry soils. Summer.

355. Fog-fruit (*Phyla nodiflora*, fig. 196). Low creeping and spreading perennial, with erect (3-5 in. high) axillary flower stalks bearing a short spike or head of tiny verbena-like purple or white flowers; corolla 2-lipped with 4 lobes, tube curved; stamens 4 included: leaf-blades broadest above the middle, coarsely toothed on end, the sides entire and tapering to the base. In very moist soil, especially near the coast. Summer.

LOPSEED FAMILY *(Phrymaceae)*

356. Lopseed *(Phryma leptostachya)*. Slender, erect perennial, 1-3 ft. high, branched above with the small, white or purplish-tinged flowers in spikes; calyx tube cylindric persisting and in fruit is bent downward and appressed against the stem, hence the common name; corolla slightly over 1/4 in. long, 2-lipped, the upper lip merely notched, the lower 3-lobed; stamens 4, the pairs unequal, included: leaves ovate to lance-shaped, toothed, opposite; stem angled and generally tinged with purple above the nodes. In woods especially in the mountain region. Spring and fall.

Mint Family *(Lamiaceae)*

KEY TO THE GENERA OF THE MINT FAMILY

A. Anther bearing stamens 2.
 1. Calyx 2-lipped.
  B. Flowers 2-4 from paired bases of reduced upper leaves (bracts).
   2. Stamens prominently exserted beyond the corolla.
    Large stone-root *(Collinsonia)*, 381.
   2. Stamens not exserted beyond the corolla. Sage *(Salvia)*, 370.
  B. Flowers many in compact axillary clusters.
   3. Leaf-blades mostly over 1¼ in. long, broadest below the middle, teeth many. False horsemint *(Blephilia)*, 322.
   3. Leaf-blades mostly under 1¼ in. long, broadest above the middle, teeth few. Pennyroyal *(Hedeoma)*, 373.
 1. Calyx not lipped, lobes about equal.
  C. Upper leaves with flowers in the axils, as large as the lower leaves or nearly so.
   4. Corollas radial. Water hoarhound *(Lycopus)*, 378.
   4. Corollas bilateral. Stone-mint *(Cunila)*, 377.
  C. Upper leaved reduced in comparison to lower leaves.
    Horse mint *(Monarda)*, 371.
A. Anther bearing stamens 4.
 5. Corollas radial or nearly so.
  D. Stamens prominently exserted beyond corolla tube; tube curved.
    Blue curls *(Trichostema)*, 359.
  D. Stamens not prominently exserted; tube straight.
   6. Flowers numerous in a series of compact clusters on upper part of stem. Mint *(Mentha)*, 379.
   6. Flowers 2-4 in upper leaf axils.
    False pennyroyal *(Isanthus)*, 358.
 5. Corollas bilateral, more or less definitely 2-lipped.
  E. Calyx with curious protuberance on upper-side.
    Skull-cap *(Scutellaria)*, 360.
  E. Calyx without protuberance.

7. Upper and lower stamen pairs of equal length.
   F. Leaves 4; flowers in a raceme.
   Small stone-root *(Micheliella)*, 380.
   F. Leaves more than 4; flowers in terminal heads or series of compact clusters. Mountain mint *(Koellia)*, 375.
7. Upper pair of stamens longer than the lower pair.
   G. Stems creeping.
      8. Flowers in a terminal 1-sided spike.
      Trailing dragon-head *(Meehania)*, 362.
      8. Flowers in axillary clusters.
      Running ivy *(Glechoma)*, 364.
   G. Stems erect.
      9. Upper lip of corolla apparently absent; stamens projecting above in position of upper lip.
      American germander *(Teucrium)*, 357.
      9. Upper lip of corolla present.
         H. Plants mostly over 3 ft. tall; spikes over 5 in. long.
         Giant hyssop *(Agastache)*, 361.
         H. Plants mostly under 3 ft. tall; spikes under 5 in. long.
         Catnep *(Nepeta)*, 363.
7. Upper pair of stamens shorter than the lower.
   I. Plants creeping. Henbit *(Lamium)*, 369.
   I. Plants erect.
      10. Calyx 2-lipped, lobes unequal.
         J. Corollas over 1 in. long, marsh plant.
         Beautiful marsh mint *(Macbridea)*, 366.
         J. Corollas under 1 in. long; not marsh plants.
            11. Upper lip of corollas strongly concave; flowers in terminal spikes. Self-heal *(Prunella)*, 365.
            11. Upper lip of corollas not concave; flower in axillary clusters. Basal *(Clinopodium)*, 374.
      10. Calyx not 2-lipped; lobes equal or nearly so.
         K. Upper lip flat, not concave.
         Bittermint *(Mesosphaerum)*, 382.
         K. Upper lip concave.
            12. Corollas about 1 in. long; calyxes faintly nerved.
            Dragon-head *(Physostegia)*, 367.
            12. Corollas ½ in. or less in length; calyxes strongly 5-10 nerved. Hedge-nettle *(Stachys)*, 368.

357. American germander *(Teucrium)*. Perennial, 1-4 ft. tall, with terminal spikes, the pink flowers blooming from be-

FIG. 197. THE MOUNTAIN MINT BEARS ITS FLOWERS IN AXILLARY CLUSTERS AS WELL AS AT THE END.

low upward; calyx bell-shaped; lower lip of corolla with 2 small side lobes, the middle one long and bends downward; stamens 4 projecting between lobes of the short upper lip; carpels 4 united below into a 4-lobed ovary and above into a 2-cleft stigma: leaves opposite, the blades ovate in *T. canadense,* toothed, pubescent. In moist ground and marsh borders, central and western. Summer.

A smaller species (2 ft. high) is found near the coast *(T. littorale)* having narrower leaves with prominent veins.

358. False pennyroyal *(Isanthus brachiatus).* Annual, 6-20 in. tall, with small, blue flowers in the upper leaf axils; calyx teeth equal, as long as the corolla tube; corolla lobes (petals) nearly equal; stamens slightly exserted: leaf-blades lance-shaped, opposite covered with a thin sticky-pubescence. In dry soils, or sandy stream borders, central and western. Summer.

359. Blue curls *(Trichostema).* Annuals, 6-22 in. high, with small, blue or white flowers borne on slender stalks from the axils of the reduced upper leaves (bracts); upper 3-lobed part of calyx much longer than the lower 2-lobed part; corolla tube slender bearing 5 narrow lobes nearly equal; stamens long and curved, very prominently exserted beyond the corolla: leaves opposite, the blades lance-shaped *(T. dichotomum)* or linear *(T. lineare).* In dry uplands. Late summer.

360. Skullcap *(Scutellaria).* Perennials, 1-3 ft. tall, with rather showy, blue, purple or white flowers in erect terminal spike-like racemes; calyx 2-lipped, bearing on the upper side the peculiar flattened pocket-like protuberance or appendage which gives the calyx the shape of the ancient death-cap, suggesting the common name; no other flower has such a calyx structure; corolla 2-lipped, the lower lip larger with 2 side-lobes: leaves opposite, the blades varying in our 10 or 12 species from nearly circular in outline to narrowly lance-shaped, rather coarsely toothed in most species, opposite. Uplands, but avoiding the drier soils. Summer.

361. Giant hyssop *(Agastache).* Tall coarse perennial, 2-6½ ft. high, with terminal compact spikes of unattractive greenish yellow or purplish flowers; calyx with the 3 upper teeth slightly longer than the 2 lower; corolla tube about as long as the calyx, bearing 2 short lips, the upper one erect, the lower bending downward, 3-lobed; stamens 4, upper pair longer than lower: leaves opposite, the blades broadly ovate, sharp-pointed coarsely toothed, smooth in *A. nepetoides* and pubescent in *A. scrophulariaefolia.* In moist ground of woodlands, chiefly in the western mountain region.

362. Trailing dragon-head *(Meehania cordata).* Perennial, 3-8 in. high, spreading by leafy runners bearing erect bracted or leafy spikes of showy blue flowers; calyx lobes equal; corolla 1 in. long, 2-lipped, upper lip 2-lobed, lower 3-lobed; stamens 4 bending upward under the upper lip:

leaves opposite, the blades heart-shape, with rounded teeth. In rich mountain woods. Summer.

363. Catnep *(Nepeta Cataria).* Erect perennial with a terminal series of leaf axil clusters of small purplish or whitish flowers. Calyx elongated, tubular, generally curved, the teeth slender; corolla strongly 2-lipped, the lower of 3 lobes, the middle one much larger with a finely wavy margin: leaves opposite, the blades ovate in outline, coarsely toothed, covered with a dense short pubescence, stalked. In waste ground, chiefly in the western part of our area. Summer and fall.

364. Ground ivy *(Glecoma hederacea).* Very low creeping plant with small, blue or white flowers in the leaf axils: corolla tube longer than the calyx teeth, 2-lipped, upper with 2 lobes, lower with 3; upper pair stamens longer than lower: leaves opposite, the blades nearly circular in outline with coarse rounded teeth. In woodlands and waste places. Spring.

365. Self-heal *(Prunella vulgaris).* Low, erect, 3-24 in. tall, herb more or less branched with terminal short compact spikes of coarse bluish-purple flowers; calyx prominently veined, strongly 2-lipped, upper lip with 3 short teeth, lower with 2 very slender sharp pointed teeth; corolla 2-lipped, the upper arching over the spreading 3-lobed lower lip; stamens 4, the lower pair longer than the upper: leaves opposite, the blades ovate or lance-shape, the lower on long stalks, mostly toothed. In open or wooded moist uplands, chiefly central and western. Spring to fall.

366. Beautiful marsh mint *(Macbridea pulchra).* Perennial, 1-2 ft. high, with rose-colored striped flowers clustered in the upper leaf axils; calyx 3-lobed; corolla about 1¼ in. long, 2-lipped, the upper of a single rounded lobe, the lower 3-lobed, the middle lobe, notched; stamens 4 with pubescent anthers: leaves opposite, the blades elliptic or nearly linear, margin wavy-toothed. In marshes, chiefly in coastal plain. Summer and fall.

367. Dragon-head *(Physostegia).* Perennials, 1-4 ft. high, with bright pink flowers in terminal showy spikes; calyx with 5 equal teeth; corolla about 1 in. long with funnel-form tube, 2-lipped, the upper one entire or notched, the lower 3-lobed; stamens 4, the lower pair longer than the upper pair: leaves opposite, the blades narrow lance-shaped, with sharp pointed teeth in *P. Virginianum* and weakly toothed in *(P. denticulata).* In marshes and very moist soil, central and western. Summer.

368. Hedge nettle *(Stachys).* Unattractive perennials, 1-4 ft. high, with the purple or white flowers clustered in the axils of the upper reduced leaves (bracts); corolla tube narrow, short (not longer than calyx) ending in 2 lips, the upper entire or notched, the lower 3-lobed; stamens 4, the upper pair longer than the lower: leaves opposite, the blades varying in our 9 species from ovate to linear *(S. hyssopifolia),* in the broader leaved species, toothed, sessile or short stalked. Most of our species prefer moist soil, chiefly western. Summer.

369. Henbit *(Lamium).* Spreading annual or biennial weed with reddish-purple flowers clustered in the upper leaf axils: sepals about equal, sharp pointed; corolla 2-lipped, the upper entire, the lower 3-lobed, the middle one spotted: leaves opposite, the blades nearly circular in outline with coarse rounded teeth, the upper ones sessile as in the henbit *(L. amplexicaule),* or short stalked as in the red dead nettle *(L. purpureum).* In waste areas and yards, adventive from Europe. Spring and summer.

370. Sage *(Salvia).* Perennials, 1-3 ft. high, bearing the purple or white flowers clustered in whorls on the upper part of the stem; calyx 2-lipped, the upper with 3 small teeth, the lower with 2 prominent teeth; corolla 2-lipped, upper entire (without notch), the lower 3-lobed; stamens 2, one cell of the anther reduced or absent: leaves opposite, the lower or basal leaf-blades deeply lobed *(S. lyrata)* or broadly ovate and toothed *(S. urticifolia)*. In rather dry uplands. Spring and summer.

371. Horse mint and others *(Monarda)*. Rather strong scented perennials, 1-3 ft. high, with the flowers in terminal head-like clusters or in axillary whorls; calyx with 5 equal teeth; corolla tube strongly 2-lipped the upper lip entire or slightly notched, the lower 3-lobed, the middle one longest: stamens 2, extending slightly beyond the upper lip: leaves opposite, the blades broadly ovate to lance-shaped, toothed. About 5 species, some in moist soil, others in dry uplands, chiefly western. Summer.

KEY TO SPECIES

Flowers scarlet. American bee balm *(M. didyma)*.
Flowers greenish or yellowish.
  Flowers in terminal head only. Basal balm *(M. Clinopodia)*.
  Flowers in axillary whorls as well as terminal.
    Common horse mint *(M. punctata)*.
Flowers purplish.
  Bracts next flowers colored with deep purple.
    Purple bergamot *(M. media)*.
  Bracts if colored nearly whitish or light purple.
    Wild bergamot *(M. fistulosa)*.

372. False horsemint *(Blephilia)*. Perennials, 1-3 ft. high, similar to the preceding genus but separated from it by the 2-lipped calyx in which the upper 3 teeth are definitely longer than the lower 2; flowers in a series of compact whorls on the upper part of the stem, purplish: leaves opposite, the blades ovate or lance-shaped, toothed, long stalked in *B. hirsuta* and short-stalked or sessile in *B. ciliata*. Two species in dry woods, western. Summer.

373. Pennyroyal *(Hedeoma pulegioides)*. Low fragrant annual, 6-18 in. high, with slender branching stem bearing compact whorls of the minute purple flowers; calyx with pubescence in the throat and slightly swollen on the under side; corolla 2-lipped, less than 1/4 in. long: leaves opposite, the blades narrowly ovate with a few teeth. In dry woods and fields, chiefly central and western. Summer.

374. Basil *(Clinopodium)*. Perennials, 1-2 ft. high, with purple or white flowers in axillary clusters; calyx teeth unequal, the 2 lower longer than the 3 upper; corolla tube straight, the upper lip notched, the lower with 3 rounded lobes; stamens 4: leaves opposite, the leaf-blades ovate, shallowly toothed. Two or three species of which the basil-weed *(C. vulgare)* with its dense clusters of many flowers, is the most common. In dry woods and fields, western. Summer.

375. Mountain mint (*Koellia,* fig. 197). Perennials, 1-3 ft. high, with numerous compact terminal head-like clusters or with dense whorls of small, white or purplish flowers, often forming in the branching species a flat-topped inflorescence; calyx teeth equal or nearly so; corolla 2-lipped, the upper entire or notched, the lower 3-lobed; stamens 4: leaves opposite, the blades varying in our 15 species from lance-shape to linear. Mostly in dry soil, but *K. aristata* and *hyssopifolia* are found in grassy bogs. Summer and fall.

376. Wild thyme *(Thymus Serpyllum)*. Creeping perennial with upright, short branches terminated by compact clusters of small, purple flowers; calyx 2-lipped, the upper 3-toothed one larger than the 3-toothed lower one; corolla 2-lipped, the upper shallowly notched, the lower 3-lobed; stamens exserted: leaf-blades narrowly ovate, opposite, sessile. Of European origin, escaping into waste areas in our western area. Summer.

377. Stone mint *(Cunila origanoides)*. Fragrant perennial, 8-20 in. high, with slender branching stem bearing few-flowered axillary clusters of pink-purple flowers; calyx equally 5-toothed, corolla upper lip notched, lower 3-lobed; stamens 2, long, straight, prominently exserted: leaves opposite, the blades broadly ovate, sharp but not deeply toothed, sessile. In dry woods, western. Late summer.

378. Water hoarhound *(Lycopus).* Perennials, 6 in.-3 ft. high, with small, white or purple flowers in numerous clusters in the axils of the upper leaves, which are reduced little if any; calyx with equal teeth; corolla tube short ending in 4 lobes, one of which is slightly larger with a minute notch commonly present; anther bearing stamens 2: leaves opposite, the blades ovate or lance-shaped mostly tapering below with a short petiole, sharply toothed, in *L. Americanus* the teeth are so deep as to give the leaf a lobed appearance, strikingly arranged in 4 rows to end of stem. In wet soil and marsh borders. Summer.

379. True mints *(Mentha).* Perennials, 1-3 ft. high, and mostly aromatic herbs with small, purple, pink or white flowers in dense clusters in the axils of much reduced leaves (bracts), the clusters often being so close together as to form a continuous spike-like inflorescence; corolla tube 4-lobed, the upper one commonly notched; stamens 4: leaves opposite, blades ovate to lance-shaped, toothed. In moist and wet soils. Summer.

KEY TO SPECIES

Stems smooth.
- Leaves mostly sessile. — Spearmint *(M. spicata).*
- Leaves with stalks.
  - Lowest clusters in axils of bracts. — Peppermint *(M. piperita).*
  - Lowest clusters in axils of leaves.
    - With terminal spike of flowers. — Bergamot mint *(M. citrata).*
    - With no terminal spike, flowers all axillary. — Creeping mint *(M. gentilis).*

Stems pubescent.
- Terminal spike present. — Round leaved mint *(M. rotundifolia).*
- Terminal spike absent; flowers axillary.
  - Leaf-blades tapering or wedge-shaped at base. — American wild mint *(M. Canadensis).*
  - Leaf-blades rounded at base. — Corn mint *(M. arvensis).*

380. Small stone-root *(Micheliella verticillata).* Coarse perennial, 5-20 in. high, with a terminal open or loose inflorescence of stalked flowers; the dull reddish corolla 2-lipped, the middle lobe of the lower lip fringed; stamens 4, well-exserted; leaves few, opposite, usually but 4, blades ovate, coarsely toothed. In rich mountain woods. Summer.

381. **Large stone-root** *(Collinsonia).* Tall, coarse perennial, 2-5 ft. high, similar to the preceding genus, the light yellow flowers, however, having but 2 stamens; lower 2 calyx teeth much longer than the upper 3; corolla tube 5-lobed, the lobes equal except the lowest which is much larger and fringed: leaves opposite, the blades ovate or elliptical, coarsely toothed, stalked. In moist woods. Summer and fall.

382. **Bitter mint** *(Mesosphaerum rugosum).* Perennial, 2-5 ft. high, with the compact clusters of small flowers aggregated into a dense inflorescence; calyx of each flower strongly pubescent; corolla 2-lipped, the upper 2-lobed, the lower 3-lobed, the middle lobe with a ragged margin: leaves opposite, the leaf-blades lance-shaped, sessile and continuing down the stem as wings. In wet soil and marshes of the eastern region. Summer and fall.

Figwort Family *(Rhinanthaceae)*

383. **Mullen** *(Verbascum).* Biennial weeds with yellow flowers in a spike or raceme: corolla lobes 5, nearly equal; stamens 5, unequal, attached at base of corolla tube: the sessile leaves alternate, dense woolly in the tall or true mullen *(V. Thapsus)* and smooth in the moth mullen *(V. Blattaria).* In waste areas, adventive from Europe.

384. **Toad-flax** *(Linaria).* Annual, biennial, or perennials, 1-3 ft. high, with erect branched stems bearing terminal showy racemes of blue *(L. Canadensis)* or the uncommon yellow *(L. Linaria)* flowers; corolla tube spurred at base, above 2-lipped, the upper 2-lobed lip covering the lower 3-lobed lip when in bud; stamens 4, included: leaves linear, abundant on basal branches of the blue toad-flax. This latter plant represents the genus in our state where in the spring it often turns old fields blue with its azure flowers. Spring and summer.

FIG. 198. THIS ROSE-COLORED TURTLE-HEAD WILL ARREST EVERYONE'S ATTENTION ON THE HIGH MOUNTAIN TRAILS.

385. Figwort *(Scrophularia).* Coarse, very tall, strong scented perennial, 3-10 ft. high, with relatively small, dull yellow or purplish flowers in terminal open or loose clusters; corolla tube short, swollen with 5 lobes, the 2 upper longer than the other 3; anther bearing stamens 4, the fifth (upper one) reduced to a scale: leaf-blades ovate or some narrower, coarsely toothed on long stalks. Two species in woods, chiefly in the mountains. Summer.

386. Turtle-head *(Chelone).* Perennials, 1-3 ft. high, with rather large (1 in. long) flowers in terminal spikes; corolla tube 2-lipped, suggesting the shape of a turtle-head; upper lip arched with margin entire or slightly notched, lower 3-lobed, pubescent within; one of the 5 stamens reduced and without an anther: leaves opposite, blades smooth, ovate to lance-shape, toothed. In very moist soil, chiefly central and western. Summer.

KEY TO SPECIES

Flowers white or cream-colored.
  Leaf-blades sessile, mountain species.
    Mountain turtle-head *(C. Cuthbertii)*.
  Leaf blades stalked. Common turtle-head *(C. glabra)*.
Flowers rose-color or pink.
  Flowers light red or pink, leaf-blades lance-shape, lowland species.
    Pink turtle-head *(C. obliqua)*.
  Flowers rose-color (purplish), leaf-blades ovate, high mountain species. Rose-colored turtle-head (*C. Lyonii*, fig. 198).

387. Beard-tongue *(Penstemon)*. Perennials, 1-3 ft. high, with terminal racemes of, or loose clusters of rather showy, purple or white flowers; corolla tube commonly straight, indefinitely 2-lipped, upper lip 2-lobed, lower 3-lobed; one stamen without an anther, the stalk heavily bearded: leaf-blades mostly lance-shape, or narrower, toothed in most of our 9 species. On dry and moist uplands, chiefly central and western. Summer.

388. Monkey-flower (*Mimulus*, fig. 199). Perennials, 1-4 ft. high, with attractive violet-colored flowers in the axils of the upper leaves; corolla tube not much longer than the calyx, and 2-ridged on the lower side within, strongly 2-lipped, the mouth nearly closed; upper lip 2-lobed, lower 3, all rounded; stamens 4: leaves opposite, blades ovate or more commonly lance-shaped, sharply toothed, smooth. In marshes and wet ground. Two species. Summer.

389. Water-hyssop *(Monniera monniera)*. Very low creeping and spreading plant with small (1/4 in. long) white or blue flowers borne singly on slender stalks, from the leaf axils; calyx with the upper tooth broadest, corolla tube ending in 5 rounded lobes nearly equal: leaves opposite, blades broadest above the middle, fleshy with insignificant teeth. On moist shores of sounds and quiet water bodies, near the coast. Summer.

390. Blue hedge hyssop *(Septilia)*. Fragrant, low, creeping plant with short upright branches bearing solitary blue.

FIG. 199. THE GRIMACING FACES OF THE MONKEY-FLOWERS LOOK OUT FROM THE MARSH BORDERS. PHOTO BY H. L. BLOMQUIST.

short-lived flowers on short stalks; sepals nearly separate, the upper one wider than the rest; corollas 2-lipped, the upper notched: leaf-blades sessile, ovate, with a few parallel veins, opposite. In wet soil of swamp borders, eastern. Summer and fall.

391. Hedge hyssop *(Mecardonia, Gratiola).* Annuals or perennials, 1-2 ft. high, with stalked, medium (1/2 in. long) white or yellow flowers borne from the upper leaf axils, sepals nearly distinct; corolla tube slightly 2-lipped, the upper lip entire or notched, the lower 3-lobed; stamens with anthers 4 *(Mecardonia),* or 2 only with anthers *(Gratiola)*: leaf-blades ovate or lance-shape, small, sessile (never over 2 in.) definitely toothed in all except the attractive golden hedge hyssop *(G. aurea)* in which they are nearly entire. About 5 species in wet soil of marsh and stream borders. Summer.

392. Hairy hedge-hyssop *(Sophronanthe pilosa).* Similar to the preceding 2 genera but distinguished from them by the short-stalked purplish (rarely white) flowers, the very rough, hairy foliage and the habitat for this species which is dry upland soil. Summer.

393. False pimpernel *(Ilysanthes).* Annuals or biennials, 4-20 in. high, with small, slender-stalked white or purple flowers from the upper leaf axils; flowers similar to those of *Gratiola* (see second genus above) but the 2 stamens without anthers have forked or 2-lobed stalks: leaf-blades, broadly ovate, sessile, weakly toothed, smooth. In wet soil. Summer.

394. Mud-flat pimpernel *(Hemianthus, Micranthemum).* Very low creeping and spreading herbs with short (1-3 in. high) upright branches bearing tiny insignificant purple or white flowers under 1/4 in. long; sepals 4 or 5 partially united; corolla 2-lipped: leaves minute, blades oval or orbicular, not over 3/8 in. long, smooth, opposite. On mud flats, chiefly near the coast. Summer and fall.

395. Speedwell *(Veronica).* Mostly low annuals or perennials, 4-10 in. high, with small, blue, purple or white flowers in terminal racemes or solitary in the upper leaf axils; corolla nearly radial, 4-lobed; sepals 4, practically distinct; corolla nearly radial, 4-lobed; stamens 2; fruit flattened, notched at the top: leaves various, opposite. About 8 species, some in moist, some dry soil areas. Spring and summer.

396. Culver's-root *(Leptandra Virginica).* Tall perennial, 3-7 ft. high, with rather showy dense slender-conical racemes of small pink or white flowers; calyx very short, 4-parted; corolla tube slender, about 1/4 in. long ending in 4 short nearly equal petals; stamens 2 prominently exserted: leaves opposite, blades lance-shaped, finely toothed. In moist soils, especially in the mountains. Summer.

397. Mullen foxglove *(Afzelia).* Rather coarse annuals or perennials, 1-5 ft. high, with numerous yellow flowers scattered over the widely branching plants in the upper leaf axils; sepals united below forming a tube; corolla funnel-form with 5 nearly equal lobes; stamens 4, exserted: leaves much divided, into fine segments *(A. cassioides)* or coarser segments with nearly entire leaves above *(A. pectinata)*, opposite. In sandy soil and savannahs of the coastal plain. Summer and fall.

398. False foxglove *(Dasystoma).* Mostly medium to tall perennials, 1-6 ft. high, with large showy yellow flowers in the axils of the upper reduced leaves often forming racemes; corolla funnel form with 5 spreading lobes; stamens 5, included, the stalks of the anthers, hairy: leaves opposite, blades toothed or cut. Five species in medium to dry soils. Summer and fall.

KEY TO SPECIES

Corolla hairy without.
- Plants covered with very sticky secretion.
  - Sticky false foxglove *(D. pectinata).*
- Plants not covered with sticky secretion.
  - Lousewort false foxglove *(D. Pedicularia).*

Corolla smooth.
- Stems pubescent. Downy false foxglove (*D. flava*, fig. 200).
- Stems smooth.
  - Leaf-blades. Small smooth foxglove *(D. laevigata).*
  - Leaf-blades toothed and cut.
    - Tall smooth false foxglove *(D. Virginica).*

FIG. 200. THE TALL YELLOW OR DOWNY FALSE FOXGLOVE BENDS WITH SUMMER BREEZES.

399. Gerardia (*Gerardia*, fig. 201). Annuals or perennials, 1-4 ft. high, with showy rose-purple flowers scattered loosely over the delicate branching plants; corolla funnelform, 5-lobed, slightly bilateral with 4 included stamens: leaves small, linear, opposite. About 10 species much alike in moist soil, savannahs and marshes. Summer and fall.

400. Eared Gerardia *(Otophylla auriculata)*. Rough plant similar to the preceding except the leaf-blades are lance-shaped with a curious appendage-like narrow lobe at the base on one side. In moist soil in the mountains. Summer.

401. Blue-hearts *(Buchnera Americana)*. Rough-surfaced perennial, 1-2½ ft. high, with long terminal spikes of blue-purple flowers; calyx tubular with very short teeth; corolla tube slender like that of phlox with 5 narrow wide spreading lobes or petals; stamens 4 included: leaves opposite, the blades lance-shaped, coarsely toothed. In moist sandy lands, central and eastern. Summer and fall.

FIG. 201. THE GERARDIAS SHOULD CERTAINLY BE INCLUDED AMONG THE "400" OF THE FLOWER SOCIETY.

402. Painted cup (*Castilleja coccinea,* fig. 94). Very attractive biennial, 4-20 in. high, with brilliant scarlet flowers and bracts borne in a terminal spike; each bract is 3-5 lobed; calyx tubular, flattened and split somewhat on the upper side and under sides giving a 2-toothed aspect; the compressed corolla tube about as long as the calyx, strongly 2-lipped, the upper lip long and arched, the lower short and 3-lobed; blades of stem leaves lobed, basal ones entire, the main veins parallel. In moist meadows of the piedmont and mountains. Early summer.

403. Cow-wheat *(Melampyrum).* Annuals, 4-16 in. high, with small (1/2 in. long) yellowish-white flowers borne in the axils of the upper reduced leaves; calyx sharply 4-toothed; corolla 2-lipped, the upper notched, the lower minutely 3-lobed, the sinuses or notches terminating 2 grooves in the tube below: leaves opposite, the blades ovate or lance-shaped, entire on short stalks. Two species much alike. In dry woods in the mountain region. Summer.

404. Lousewort *(Pedicularis)*. Perennials, 1/3-3 ft. high, of coarse aspect with short terminal spikes of dull, yellow flowers; calyx short, 5-toothed; corolla strongly 2-lipped, the upper lip arched over the lower 3-lobed one; stamens 4, the stalks curving into the upper lip: leaves alternate, the blades much toothed or cut, more or less hairy. In dry woods is found the low common *P. Canadensis* while the taller *P. lanceolata* is confined to marshes. Summer.

405. Chaff-seed *(Schwalbea Americana)*. Perennial, 1-2 ft. high, with open spikes of long (1-1½ in.) slender yellowish-purple flowers; calyx tube shortened on upper side with a small tooth, the other 4 teeth longer; corolla tube, narrow, cylinder-like, strongly 2-lipped, the upper arched and concave, the lower shorter and 3-lobed; stamens 4, curving into the upper corolla lip: leaf-blades mostly ovate, entire, 5-veined, sessile. In moist sandy soil, near the coast. Spring and summer.

Ruellia Family *(Acanthaceae)*

406. Sandhill twin-flower *(Calophanes oblongifolia)*. Perennial, 4-12 in. high, with attractive blue flowers spotted with purple, borne single in the upper leaf axils; calyx hairy, corolla tube broadly funnel form, the rounded 5 petals spreading above; stamens 4, included: leaves opposite, the blades ovate or nearly so, entire, upper ones sessile. In dry sandhills near the coast. Spring and summer.

407. Twin blue-bells *(Ruellia)*. Perennials, 1-4 ft. high, with large blue flowers sessile in the upper opposite leaf axils, thus forming pairs or twins; corolla tube slender in lower half and the funnel form above bearing wide flaring petals; stamens 4: leaves opposite, the blades ovate or lance-shape, sessile, entire. Three species, much alike. In dry woods. Summer.

408. False mint *(Diapedium brachiatum).* Probably annual, 1-2 ft. high, with the aspect of a mint bearing short stalked clusters of purple flowers in the upper leaf axils; corolla ¾ in. long, cleft into 2 prominent lips, the upper minutely 3-toothed, the lower entire; stamens 2; ovary 2-celled: leaves opposite, blades ovate or lance-shaped with tapering tips, short stalked, entire. In medium or dry uplands. Summer.

409. Water willow *(Dianthera).* Perennial, 4 in.-3 ft. high, with long stalked clusters of purplish or nearly white flowers borne from the upper leaf axils; corolla about 3/8 in. long, 2-lipped, the upper entire or slightly notched, the lower deeply 3-cleft; stamens 2; ovary 2-celled: leaves opposite, the blades lance-shaped or narrower, sessile, entire. Two species in running water or wet soil, central and eastern. Summer.

BLADDERWORT FAMILY *(Pinguiculaceae)*

410. Bog false violet and **butterwort** *(Pinguicula).* Low insectivorous herbs, 2-8 in. high, with showy flowers borne solitary on the end of the single erect stalk; calyx 2-lipped; corolla 5-lobed, more or less 2-lipped, the tube spurred; stamens 2; leaves forming basal rosette, slimy margins inrolling catching small insects. In grassy bogs of savannahs, eastern. Spring.

KEY TO SPECIES

Corolla violet or sometimes white. Bog false violet *(P. elatior).*
Corolla yellow. Butterwort *(P. lutea).*

411. Bladderwort *(Utricularia).* Aquatic and bog plants with chiefly yellow flowers (2 species violet-color) borne in a simple few-flowered raceme above the water or wet soil surface; calyx 2-lobed; corolla strongly 2-lipped, the throat closed by a 2-lobed pocket-like upgrowth from the lower lip (palate); stamens 2; leaves finely divided into numerous thread-form lobes or in the bog species so reduced as to be unnoticed; in the aquatic species they bear the tiny bladders which are equipped with trapdoors to catch minute water animals. Eleven species in quiet waters and on bogs, central and eastern.

Broom-rape Family *(Orobanchaceae)*

412. Squaw-root *(Conopholis Americana).* Curious low fleshy perennial, of a brownish color, not over 10 in. high, with flowers in a compact bracted spike; calyx split beneath, corolla 2-lipped, the upper concave and arched, the lower 3-lobed; stamens 4, slightly exserted: leaves light brown reduced to over-lapping scales becoming smaller toward the base. A parasitic plant of rich woods. Spring and summer.

413. Broom-rape *(Orobanche minor).* Yellow-brown plant with a broken spike of yellow flowers; calyx deeply 4-cleft; corolla 2-lipped, the upper somewhat incurved, the lower 3-lobed about 1/2 in. long; stamens 4, included: leaves reduced to scales, the longest about 3/4 in. becoming smaller below. Parasitic on clover roots. Spring and summer.

414. Naked broom-rape *(Thalesia uniflora).* Low, light colored herbs with short scaly stems close to the ground bearing long (3-8 in.) slender flower stalks, each terminated by a single white or violet colored and scented flower; calyx 5-cleft; corolla slightly 2-lipped, the lobes being nearly alike: leaves few basal, reduced to pinkish scales. Parasitic on roots in woods, chiefly western. Spring and summer.

415. Beech-drops *(Leptamnium Virginianum).* Low purplish brown plants, 6-24 in. high, branched above, the branches ending in open spikes of pale flowers; calyx with 5 short teeth; corolla 4-lobed, the upper one slightly larger and arching; stamens 4, included: leaves reduced to insignificant scales. Parasitic on beech roots. Summer.

Unicorn-plant Family *(Martyniaceae)*

416. Unicorn-plant *Martynia Louisiana).* Annual, 1-3 ft. high, with widely spreading branches bearing large yellow flowers mottled with purple, 1-1½ in. long, in short racemes; calyx tube swollen; corolla tube curved, the lobes nearly equal; stamens 4, exserted: leaf-blades nearly circular in outline, heart shaped at base, the whole plant sticky pubescent. Each fruit capsule remarkable for its large curved hook, the whole being 4-6 in. long. An escaped plant, a native of western United States. In dry uplands. Summer.

SANDALWOOD FAMILY *(Santalaceae)*

417. Bastard toad-flax *(Comandra umbellata).* Smooth perennial, 6-18 in. tall, with greenish-white flowers in many clusters, forming a conical or rounded inflorescence; "corolla" (really the calyx) tube short, 5-lobed, attached below to the ovary; stamens 5: leaf-blades narrowly ovate or lance-shaped, sessile, alternate. Parasitic on roots of other plants in dry uplands. Spring and summer.

MADDER FAMILY *(Rubiaceae)*

418. Clustered bluets *(Oldenlandia uniflora).* Low weak stemmed herbs, 4-5 in. long, bearing one or a few white flowers sessile in the upper leaf axils; flowers small, less than 1/8 in. broad, corolla tube 4-lobed, stamens 4, included: leaf-blades oval, sessile, opposite, not over 1 in. long. In sandy bogs, eastern. Summer.

419. Bluets *(Houstonia).* Low-herbs, mostly perennials, under 18 in., with attractive small, blue, purple or white flowers borne singly or in few-flowered terminal clusters; corolla tube slender, the 4 petals spreading: stamens 4, included: leaf-blades linear to ovate, sessile, opposite. On dry uplands and openings. Spring and summer.

KEY TO SPECIES

Flowers long stalked borne singly in the upper leaf axils.
  Flowering stems arising from creeping leafy stems.
    Thyme-leaved bluet *(H. serpyllifolia).*
  Flowering stems arising from basal rosettes of leaves.

FIG. 202. A TITLE IS NOT NEEDED FOR THIS PICTURE SO FAMILIAR ARE THE BLUETS TO EVERYONE IN THE SPRING.

Flowers light blue or white. Perennial.
Common bluet (*H. caerulea*, fig. 202).
Flowers dark blue or violet. Annual. Small bluet (*H. patens*).
Flowers borne in clusters on stalks from the upper leaf axils.
Leaf-blades linear. Slender leaved bluet (*H. tenuifolia*).
Leaf-blades lanceolate. Long leaved bluet (*H. longifolia*).
Leaf-blades ovate. Large leaved bluet (*H. purpurea*).

420. Partridge berry *(Mitchella repens)*. Low, creeping perennial with white flowers borne in pairs at the stem ends; corolla tube slender, nearly 1/2 in. long, the 4 lobes recurved: stamens 4, sometimes exserted, the single red berry-like fruit is composed of the united ovaries of the 2 flowers: leaves opposite, blades nearly circular in outline, shining. In shady woods. Spring to fall.

421. Rough button-weed *(Diodia)*. Low, creeping and spreading with ascending stem tips bearing small white or pink flowers borne singly and sessile in the axils of the upper leaves; petals 4: stamens four, included: leaves opposite, blades linear to ovate, rough pubescent. Three species, one, an annual *(D. teres)* with narrow leaves in dry soil, the other two, perennial, with broader leaves are in moist soil. Spring to fall.

422. Smooth button-weed *(Spermacoce glabra)*. Similar in aspect to the preceding but differs in that the very small flowers are clustered in the leaf axils and the plant is smooth. The leaves are of the broad type, lance-shaped or ovate. In moist soil especially on river flood-plains. Summer.

423. Bedstraw *(Galium)*. Spreading or erect, mostly perennials, 6 in.- 4 ft. high or long, with tiny white or purplish flowers in terminal open clusters; corolla tube short, 4-lobed; anthers 4, exserted: stem square (4-angled) smooth or in some species covered with short recurved bristles: leaves 3 or more whorled, blades linear to broadly ovate. About 10 species occupying a wide range of habitats. Summer.

Honeysuckle Family *(Caprifoliaceae)*

424. Horse gentian *(Triosteum)*. Medium, erect perennials, 2-4 ft. tall, with the reddish and purplish brown or yellowish sessile flowers solitary or a few clustered in the upper leaf axils; sepals 5, slender longer than the ovary below them; corolla narrowly funnel-form, 5-lobed; stamens 5, stalks very short: leaves opposite, the blades large 4-10 in. long, lance-shaped or ovate, more or less contracted at the base. In rich woods mostly in the mountains. Summer.

KEY TO SPECIES

Flowers yellow. Narrow-leaved horse gentian *(T. angustifolium)*.
Flowers purplish or reddish brown.
  Leaf pairs connected by their bases.
    Orange-fruited horse gentian *(T. perfoliatum)*.
  Leaf pairs never connected by the bases.
    Scarlet-fruited horse gentian *(T. aurantiacum)*.

VALERIAN FAMILY *(Valerianaceae)*

425. Corn salad *(Valerianella radiata)*. Low, erect, rather delicate or watery annual, 6-18 in. high, with tiny white flowers borne in terminal flat-topped clusters at the ends of the regularly forking stems; corolla tube, short, 5-lobed; stamens 3: leaves opposite, the blades lance-shaped, smooth, sessile. In moist places. Summer.

TEASEL FAMILY *(Morinaceae)*

426. Teasel *(Dipsacus sylvestris)*. Biennial, 3-6 ft. tall, the few branches terminated by large conspicuous heads of lilac-colored flowers surrounded by long upward curving bracts; heads globe-shaped or longer than broad; corolla 4-lobed somewhat bilateral; stamens 4; each flower in the axil of a stiff pointed persistent scale or bract: leaves opposite, the blades lance-shaped or broader, lower with small rounded teeth, sessile. A weed from Europe chiefly occurring in the western region.

WILD GINGER FAMILY *(Asaraceae)*

427. Wild ginger *(Hexastylis* and *Asarum)*. Low perennials, under 12 in. with the short-stemmed flowers borne from the underground stem; they are commonly hidden amid the leaf mold; flowers with calyx only, 3-lobed, of a dull purplish color, sometimes mottled; stamens 12: leaves all basal, the blades varying from heart-shape to orbicular, but all have a deep heart-shaped base, mottled in some species. In rich woods. Spring and early summer.

KEY TO SPECIES

Calyx tube attached below the ovary *(Hexastylis)*.
  Leaf-blades broadly arrow-shaped.
    Calyx tubes contracted above the middle.
      Common arrow-leaved wild ginger *(H. arifolium)*.
    Calyx tubes not prominently contracted above the middle.
      Mountain arrow-leaved wild ginger *(H. Ruthii)*.

FIG. 203. THE VEINY LEAVES OF THE VIRGINIA WILD GINGER ARE DISTINCTIVE. NOTE FLOWER BELOW SMALL MIDDLE LEAF. PHOTO BY H. L. BLOMQUIST.

Leaf-blades broadly ovate or orbicular.
Four species, much alike of which *H. Virginica* (Fig. 203) is most common.
Calyx tubes attached above the ovary *(Asarum)*.
Sepals longer than ovary. Indian wild ginger *(A. Canadense)*.
Sepals shorter than ovary.
Small-flowered Indian ginger *(A. reflexum)*.

BELLFLOWER FAMILY *(Campanulaceae)*

428. Bellflower *(Campanula)*. Perennials, 1-3 ft. tall, bearing small, attractive blue or white flowers, few or solitary on the delicate upper branches; corolla funnel or bell form, 5-lobed; stamens 5, free from the corolla: leaves linear or lance-shaped, toothed, alternate. Two species, one with linear leaves *(C. aparinoides)* is found in grassy, wet places, the other with lance-shaped leaves occurs in rock crevices. Both are chiefly distributed in the mountain region. Summer.

429. Tall or American bellflower *(Campanulastrum Americana)*. Tall perennial, 2-5 ft. high, with a long terminal leafy spike of blue (occasionally nearly white) flowers; corolla tube short, the 5 lobes wide spreading and ending in a thickened tip; stamens 5 exserted: leaves alternate, the blades mostly lance-shaped, 3-6 in. long, toothed, narrowed at base. In moist woods, most common in the mountains. Summer.

430. Venus' looking-glass *(Specularia)*. Slender, erect annuals, 6-24 in. high, bearing one or a few blue flowers sessile in the leaf axils; corolla tube short, prominently 5-lobed, the lobes spreading; stamens 5, the stalks flat: the sessile leaf-blades in the small species *(S. biflora)* are ovate or lance-shaped while in the common *S. perfoliata* they are orbicular. In dry soil. Spring and summer.

Lobelia Family *(Lobeliaceae)*

431. Lobelia *(Lobelia)*. Annuals, biennials, or perennials, 6 in.-4½ ft. high, with red, blue, or white flowers in spike-like terminal racemes; the 2-lipped corolla tube is distinctive by being split to the base on the upper side; lobes 5, 2 above, 3 below; stamens 5 often projecting through corolla tube slit: leaves alternate, the blades ranging from linear to ovate, mostly toothed. Chiefly in marshes, bogs or moist soil of uplands. Summer.

KEY TO SPECIES

Flowers scarlet. Cardinal flower (*L. cardinalis,* fig. 204).
Flowers blue, blue-purplish, or white.
  Corolla over ⅜ in. long.
    Leaf-blades very narrow, linear.
      Narrow-leaved marsh lobelia *(L. glandulosa)*.
    Leaf-blades of a broader type.
      Leaves covered with dense fine pubescence.
        Upland hairy lobelia *(L. puberula)*.
      Leaves smooth or nearly so.
        Calyx with reflexed appendages between the sepals.
          Great lobelia *(L. syphilitica)*.
        Calyx without reflexed appendages.
          A group of 5 large flowered species not necessary to be distinguished here.
  Corolla under ⅜ in. Small flowered lobelias.
    A group of 7 species, mostly delicate with narrow-leaves found in bogs and dry uplands.

FIG. 204. THE RED-BIRD OF THE PLANT WORLD IS THE CARDINAL FLOWER. PHOTO BY H. L. BLOMQUIST.

THISTLE FAMILY *(Carduaceae)*

432. **Ironweed** *(Vernonia)*. Weed-like perennials, 2-10 ft. high, with numerous heads of purple flowers borne in a more or less flat-topped open cluster; each head surrounded by numerous overlapping bracts arranged in many series; flowers

FIG. 205. PURPLE HEAD WATER WEED IS ONE OF OUR MOST INTERESTING AQUATIC COMPOSITES.

all alike, radiate, the calyx consisting of 2 rows of capillary bristles, or in some species the outer row is reduced to scales: leaves alternate, blades varying around lance-shape, toothed. About 4 species in upland open places, chiefly in the western area. Late summer.

433. Elephant's foot *(Elephantopus).* Perennials, 1-3 ft. high, with the few purple flowered heads borne at the ends of forking branches; each head in addition to the small bracts is surrounded by 3 prominent leaf-like bracts, a distinctive character: leaves chiefly basal, blades varying around ovate, shallowly toothed. Three species, much alike. In dry woods. Late summer.

434. Purple-head water-weed (*Sclerolepis uniflora,* fig. 205.) Delicate spreading perennial with single heads of small purple flowers borne at the ends of the vertical branches; calyx of 5 dull pointed scales: leaves 4-6 at a joint, linear, with a single vein down the middle. In pools and slow streams of the coastal plain. Spring and summer.

435. Thoroughwort *(Eupatorium).* Perennials, 1-10 ft. high, with numerous small heads of white or purple flowers disposed in a terminal, commonly flat-topped inflorescence; each head is few flowered, rather slender, the bracts of different length, overlapping; calyx a single row of fine bristles: leaves varying widely in the different species both in shape and arrangement. Our 20 species are distributed in a wide range of habitats. Summer and fall.

Deserving especial mention are: The tall Joe-Pye weed *(E. purpureum)* with its light purple flowers and whorled leaves, the southern dog-fennel *(E. capillifolium)* with its fine thread-like leaves giving the plant a very beautiful plume-like aspect; the round-leaved thoroughwort *(E. rotundifolium)* so common in the semi-bogs of the open grassy type; the white-snake-root *(E. urticaefolium)* of the mountains with its coarsely toothed stalked leaves, the cause of "trembles" poisoning in sheep.

436. Vanilla plant *(Trilisa).* Erect perennials, 1½-4 ft. high bearing numerous heads of purplish or more rarely white flowers in a loose terminal cluster; bracts surrounding each head of nearly the same length; calyx of 1 or 2 rows of barbed bristles: stem wand-like with the largest leaves at or near the base: leaf-blades lance-shaped or ovate, the largest tapering below. In grassy bog areas, eastern. Late summer and fall.

KEY TO SPECIES

Stems smooth. True vanilla plant or hound-tongue *(T. odoratissima).*
Stems sticky, hairy. False hound tongue *(T. paniculata).*

437. Blazing star *(Laciniaria).* Attractive perennials, 1-6 ft. high, with the purple flowered heads mostly in slender terminal spikes or racemes; outer bracts of head shorter than inner ones, middle ones intergrading between; flowers attached to a flat receptacle: leaves alternate, the blades linear or nearly so, numerous; stems arise from a hard underground globular tuber. About 10 species in a wide range of soils. Summer.

Of especial interest are: The scaly blazing star *(L. squarrosa)* with its broad bracts tapering into a slender spine-like point; the common dotted blazing star with the leaves covered with minute-glands; the savannah blazing star *(L. Wellsii)* a species recently discovered having the upper part of the outer bracts colored like the flowers.

438. False blazing star *(Carphephorus).* Perennials, 1-3 ft. high, with purple heads much like those of the preceding genus borne in an open more or less flat-topped spreading inflorescence: the basal leaf-blades are definitely larger than the stem ones and are broader above the middle with rounded tips, tapering below into petioles. Three species, all on moist sandy soil or semi-bogs, chiefly in the coastal plain. Summer and fall.

439. False boneset *(Kuhnia eupatorioides).* Perennial, 1-3 ft. high, with slender purplish or white heads much resembling the true boneset *(Eupatorium)* from which it is separated by the insignificant character of number of ridges on the achenes (seeds). In true boneset the number is 5, in false, 10: leaf-blades narrowly lance-shaped, sparingly toothed. On dry slopes, chiefly in the western area. Late summer.

440. Golden aster *(Chrysopsis).* Showy biennials or perennials, 1-3 ft. high, with the few yellow-flowered heads borne loosely in a more or less flat-topped inflorescence; calyx of both bristles and scales; "seeds" (achenes) flattened; rays elliptical, showy: leaves alternate, the blades linear to lance-shaped, mostly hairy. Five species in dry woods and open places. Summer.

Especially common are: The grass-leaved golden aster *(C. graminifolia)* with its silvery grass-like leaves; the Maryland golden aster with broader lance-shaped leaves hairy when young but nearly smooth in maturity; the woolly golden aster of the dry sandy soils in the coastal plain with its leaves covered with a heavy coat of pubescence.

441. Rayless flat-topped golden rod *(Chondrophora).* Perennials, 8-24 in. high bearing numerous small few-flowered yellow heads in a definitely flat-topped inflorescence; ray flowers absent; calyx only of capillary bristles; stem leaves linear *(C. nudata)* or thread-form *(C. virgata),* the larger leaves are, however, mainly basal and slightly broader above the middle. On savannahs or grassy bogs in the coastal plain. Fall.

442. Sandhill yellow aster *(Isopappus divaricatus).* Annual or biennial, 1-4 ft. tall, bearing numerous rather showy yellow heads in a very loose or open terminal inflorescence; ray flowers not over 12; calyx or single row of bristles of equal length: leaves alternate, the blades narrow, nearly linear with a few sharp teeth. Weed-like biennial plants especially common in old fields in the sandhills. Late summer and fall.

443. Flat-topped golden-rod *(Euthamia).* Perennials, 2-4 ft. high, with numerous small yellow heads in a spreading flat-topped inflorescence; ray flowers commonly over 10; central or tubular flowers fewer; calyx of long bristles of equal length: leaf-blades linear or nearly so, entire. Two species one *(E. graminifolia)* with 3-veined leaf-blades, the other *(E. Caroliniana)* with 1-veined blades. In upland habitats, especially old fields. Summer and fall.

444. Golden-rod *(Solidago).* An extraordinarily large genus (over 40 species in North Carolina) of perennial herbs, 1-8 ft. high, bearing the numerous small, yellow-flowered heads in narrow or broad usually curved compact clusters; outer bracts of the head becoming shorter, often extending down the flower stalk; central or tubular flowers more numerous than the marginal or ray flowers: calyx of numerous bristles in 1 or 2 rows: leaves various. Nearly all of our habitats show one or more species. Late summer and fall.

A few species of especial interest are: The white golden rod *(S. bicolor)* which has dull white flowers; the sweet golden-rod *(S. odora)* which has the scent of anise; the large coarse pine barren golden rod so common on burned over peatty areas in the coastal plain and the salt marsh golden-rod *(S. sempervirens)*.

445. False golden-rod *(Brachychaeta sphacelata)*. A yellow-flowered perennial closely resembling the golden-rods, but the calyx bristles are shorter than the seed (achene) below them; the lower coarsely toothed leaf-blades are stalked and are heart-shaped at the base. In dry woods especially in the mountain region. Late summer.

446. Bolton's aster *(Boltonia asteroides)*. Medium to tall perennial, 3-8 ft. high, with rather showy purplish or white heads about 1 in. in dia. borne on a very loose or open branching system above; bracts of the hemispherical base of the head with thin dry margins; calyx of short scales with 2 to 4 rigid bristles open present; seeds (achenes) flattened, the margins winged: leaves alternate, the blades lance-shaped or inverted lance-shaped, entire, sessile. In damp soil of low grounds. Summer.

447. White-topped aster *(Sericocarpus)*. Perennials, 1-3 ft. tall, with about 5-rayed heads, borne in a definitely flat-topped inflorescence; ray flowers white, central tubular flowers, yellowish or purplish; calyx of fine bristles: leaves alternate, the blades in our 3 species varying from linear *(S. linifolius)* to the broader and entire leaved *S. bifoliatus* and the taller lance-shaped and toothed *S. asteroides.* In dry upland sunny areas. Summer.

448. Aster *(Aster)*. The genus of flowering herbs in North Carolina having the largest number of species, for no less than 60 occur within the state. Perennials, 1-8 ft. high with the purple, light purplish or white rayed heads borne in open or sometimes rather compact clusters above; rays usually numerous, over 10; central or tubular flowers usually yellow in color changing in age to reddish brown or purplish; calyx bristles mostly in a simple series about the corolla: leaves alternate, various. Nearly all of our habitats have aster representatives. The blooming period for the genus is late summer and fall.

Only a few species need be given special mention: The beautiful vigorous New England aster *(A. Novi-Angliae)* with its 2 in. broad heads; the old field aster *(A. ericoides)* so familiar on abandoned land with its small very numerous white flowers; the grass-leaved bog or savannah aster *(A. paludosus)* and the two salt marsh asters *A. tenuifolius* and *A. subulatus.*

449. Fleabane *(Erigeron).* Weed-like plants, 1-3½ ft. tall, bearing loose terminal clusters of rather showy white, pink, or purple many-rayed (commonly over 40) heads on long stalks; bracts in only 1 or 2 series, the lips reflexed; central tubular flowers with yellow corollas; calyx bristles mostly in a single series: leaves narrow, alternate. In dry upland soils or in the case of the eastern *E. vernus,* distinguished by its chiefly basal leaves, the bog or savannah soil is preferred.

450. Horseweed *(Leptilon canadense).* Annual or biennial weed, 2 in.-10 ft. high, bearing the numerous small very short white-rayed heads in a loose terminal cluster: leaves numerous, alternate, the upper ones linear and curving outwardly at right angles to the stem, pubescent with rows of hairs on the margins. In old fields, even those of most sterile soils; also on sand flats along the sounds (Fig. 206). Summer and fall.

451. False flat-topped aster *(Doellingeria).* Perennials, 1-8 ft. high, with 10-15 white rayed heads arranged in a flattish or rounded inflorescence above; tubular or central flowers having very slender corolla tubes below; calyx bristles in two series, an inner of long delicate ones and outer of short bristles or scales: leaves alternate, reduced toward the base, the

FIG. 206. INCHES HIGH INSTEAD OF FEET, THE COMMON HORSEWEED IN STERILE SEASIDE SAND IS HARDLY RECOGNIZABLE.

blades lance-shaped, sessile, entire. Three species much alike, chiefly on moist uplands. Late summer.

452. Stiff-leaved aster *(Ionactis linariifolia)*. Perennial, 6-24 in. high, with showy aster-like heads borne in an umbel-like inflorescence at the ends of slender stalks from the upper part of the stem; bracts of the head-thickened; calyx bristles in 2 series, the inner long and the outer very short: leaves alternate, the blades linear, under 1½ in. long, rigid and spreading at right angles to the stem, rough-pubescent. In dry and rocky soil, chiefly in the western area. Late summer.

453. Marsh fleabane *(Pluchea)*. Annuals or perennials, 1½-3 ft. high, with rather small heads of restricted number in a terminal cluster; calyx bristles in a single series; seeds developed only by the outer flowers: leaves alternate, the blades lance-shaped but with heart-shaped base in the perennial marsh species *(P. foetida)*, coarsely toothed. The 2 annuals are in marshes, salt marshes *(P. camphorata)* and in moist woods *(P. petiolata)*.

454. Black-root *(Chaenolobus undulatus)*. Biennial or perennial, woolly herbs, 18-40 in. high, appearing like a giant rabbit tobacco plant with sessile and densely crowded heads borne in a spike-like inflorescence above: leaves alternate, the blades lance-shaped but with base not only attached to the stem but continuing down it as two wings. In sandy pine lands of the coastal plain. Spring and summer.

454. Everlasting *(Antennaria)*. Low woolly perennials, 2-20 in. high, with the heads solitary *(A. solitaria)* compactly clustered above *(A. plantaginifolia)*; bracts of head thin, membranous; leaves chiefly basal, forming a rosette. In dry soil, preferring open woodlands. Spring and summer.

456. Pearly everlasting *(Anaphalis margaritacea)*. Perennial, 1-3 ft. high, with numerous rayless heads in a terminal flat-topped or slightly rounded cluster; bracts of heads dry, white, persistent: leaves numerous, alternate, linear, pubescent above but strongly woolly beneath. In rather dry waste places; a domestic immigrant from the north and west. Summer and fall.

457. Cudweed or rabbit tobacco *(Gnaphalium)*. Low to medium woolly herbs, 3-36 in. high, with heads in a loose flat-topped or rounded cluster or in a spike-like inflorescence *(G. purpureum)*: leaves not chiefly basal, the blades varying from lance-shaped or broader ones at base to nearly linear on the stem, commonly woolly beneath, smooth above. Three species of which the tall *G. obtusifolium* with its loose rounded cluster of heads is the most striking. In dry soils. Summer.

458. Leaf-cup *(Polymnia)*. Very coarse, tall perennial, 2-10 ft. high, bearing the white or yellow rayed heads in a branching open cluster above; bracts of 2 kinds, the outer, about 5 in number, large and conspicuous, the inner small and numerous; calyx bristles absent; central flowers not fruit pro-

FIG. 207. THE HEADS OF THE SPRING GREEN AND GOLD RESEMBLE SINGLE FLOWERS, BUT SHARP EYES WILL NOT BE CAUGHT BY THIS COMPOSITE INFLORESCENCE TRICK. PHOTO BY H. L. BLOMQUIST.

ducing: leaves large, opposite, the blades coarsely toothed, and lobed. In rich soil of woods. Summer.

KEY TO SPECIES

Rays yellow. *P. Uvedalia.*
Rays white.
  Rays very short, under ⅛ in. long. *P. Canadensis.*
  Rays longer, over ¼ in. long. *P. radiata.*

459. Rosin-weed *(Silphium)*. Coarse, tall perennials, 3-10 ft. high, with yellow-rayed showy heads borne in a loose branching cluster; bracts large, spreading at least at the tips; central flowers not fruit producing; seeds flattened; ray flowers in more than 1 series: leaf-blades mostly ovate or lance-shaped, shallowly toothed, but in *S. pinnatifidum,* the large basal ones are deeply lobed. In moist and dry soil. Summer.

460. Spring green and gold (*Chrysogonum Virginianum,* fig. 207). Low, attractive perennial, 4-12 in. high with stalked yellow-rayed heads borne solitary in the leaf axils or terminating the stems; bracts in 2 series, of 5 each, the outer 5, large, the inner small; opposite to the inner bracts are the 5 ray flowers with their broad, nearly orbicular rays; seeds flattened; leaves opposite, the blades ovate, or nearly so, toothed, softly pubescent, stalked. One of our earliest blooming composite plants. In dry woodlands. Spring and early summer.

461. Woolly leaf *(Berlandiera pumila).* Perennial, 8-36 in. tall, with the few showy yellow-rayed heads borne in small groups at the stem ends; bracts overlapping, the outer ones the smallest; ray flowers in but one series: leaves alternate, the blades ovate or nearly so, toothed with rounded teeth, covered below with a gray, woolly pubescence, the lower blades are stalked, the upper sessile. In sandy soils, eastern. Summer.

462. Fever-few *(Parthenium integrifolium).* Coarse perennial, 1-4 ft. high, bearing the numerous small heads in a terminal flat-topped inflorescence; bracts few, oval, overlapping; ray flowers 5 with very small, white, thick corollas (rays), fruit producing; the seeds (achenes) flattened; central flowers numerous, sterile: leaves alternate, the blades varying around ovate, tapering below, coarsely toothed, rough on both sides. In dry upland soils. Summer.

463. Smooth false sunflower *(Heliopsis helianthoides).* Tall, smooth perennial, 3-5 ft. high, bearing large (1½-2½ in. dia.) showy yellow-rayed heads on slender stalks from the upper leaf axils and stem ends; the ray flowers in age tend to persist on to the short, thick seeds (achenes) below; central, tubular flowers enclosed by the chaffy scales, seed producing: leaves opposite, the blades lance-shaped to ovate, toothed. In dry woods, chiefly western. Summer.

464. Hairy false sunflower *(Tetragonotheca helianthoides).* Sticky-hairy perennial, 1-2½ ft. tall, with very showy (1½-3 in. wide) long stalked heads borne on the stem ends; outer bracts 4, large, united at the base, inner small, inconspicuous; central flowers seed producing, the seeds 4-sided; corolla (ray) of the ray flowers 3-lobed at apex: leaves opposite, the blades ovate, wavy-toothed, sessile. In dry soil, chiefly in the mountain region. Early summer.

465. Yerba de tajo *(Eclipta alba).* Annual weed, 1-3 ft. tall, with flowers in inconspicuous heads; rays flowers white, short; bracts of heads in 2 series: chaffy scales among the central flowers: leaves opposite, the blades lance-shaped or narrower, few-toothed, pubescent. In moist, waste areas. Adventive from tropical America. Summer and fall.

466. False cone flower *(Spilanthes repens).* Perennial, 8-24 in. tall, bearing slender-stalked yellow-rayed heads from the upper leaf axils; bracts in 2 series, rather loosely overlapping; central flowers in the form of a narrow cone 3/8 in. high; heads slightly under 1 in. broad: leaves opposite, the blades ovate or lance-shaped; coarsely toothed, stalked. In moist soil, chiefly central. Summer.

467. Cone-flower *(Rudbeckia).* About 10 species of medium to tall stout or rigid perennials, 2-12 ft. high, with showy heads on long stalks from the upper leaf axils; rays yellow, central conical mass of flowers dark purplish; seeds produced by the flowers of the central cone, 4-angled, the rough leaf-blades of all our species except one around lance-shape, *R. laciniata* related to the golden-glow of the gardens has deeply lobed and divided blades. The thick leaved, showy species, under 3 ft. with orange-yellow rays is the rather common "black-eyed susan." Chiefly in moist soil. Summer.

468. Purple-cone-flower *(Brauneria purpurea).* Very showy perennial, 2-5 ft. tall, bearing solitary, long-stalked reddish-purple rayed heads, 3-4 in. dia.; chaffy scales amid central flowers, each keeled, stiff-pointed: leaf-blades ovate or narrower, coarsely but rather shallowly toothed, the stem leaves mostly with 3 principal veins. In moist areas chiefly in the western region. Summer and fall.

469. Sea oxeye *(Borrichia frutescens).* Perennial spreading plant, 1-2½ ft. high, with slightly fleshy stems above, woody below, bearing showy yellow-rayed heads, 1 in. or more in dia.; central flowers each in axil of a stiff bracelet or chaffy scale: leaves opposite, the blades narrow but broadest above the middle, fleshy. Forming extensive societies in salt marshes. Summer and fall.

470. Sunflower *(Helianthus).* About 18 species of medium to tall perennials, 2-7 ft. high with showy yellow-rayed heads borne solitary or in terminal loose clusters; bracts many overlapping; seeds but slightly flattened, not winged; central flowers or tubular flowers generally not yellow but purplish or brown: leaf-blades mostly of the ovate or lance-shaped type, toothed, leaves often alternate and opposite on the same plant. Mostly in dry soil, a few in bogs or moist soil. Summer.

471. Wing stem *(Actinomeris alternifolia).* Tall perennial, 4-9 ft. high, with showy stalked heads closely resembling those of the sunflowers, forming a loose terminal cluster; seeds flattened with a winged margin: leaves alternate, the blades lance-shaped, tapering below into a winged stalk the wings of which usually continue down the stem, furnishing a very distinctive character separating this species from the sunflowers. In rich, moist soils. Late summer.

472. Crownbeard *(Verbesina).* About 3 species of tall perennials, 2-7 ft. high, with terminal loose clusters of yellow or white-rayed heads; rays few under 7 except in the sunflower crownbeard which has from 8-15; bracts of involucre numerous, overlapping; seeds flattened, with wing margin: leaf-blades ovate to lance-shaped, toothed. In dry upland soils. Summer.

KEY TO SPECIES

Ray flowers white. White crownbeard *(V. Virginica).*

Ray flowers yellow.

Ray flowers about ½ in. long. Small yellow crownbeard *(V. occidentalis).*

Ray flowers about 1 in. long. Large yellow crownbeard *(V. helianthoides).*

FIG. 208. THE CURIOUS MARSHALLIA OF THE COMPOSITE FAMILY WHICH RESEMBLES A CLUSTER OF LOUDSPEAKERS.

473. Tickseed *(Coreopsis).* About 12 species of showy mostly perennial herbs, 1-4 ft. high, with long stalked heads terminating the branches; bracts in 2 very distinct series united at the base—an outer whorl of narrow ones spreading from near the base at the head, the inner ones broad, thin and appressed tightly against the included flowers; calyx mostly of 2 short teeth and persisting on the mature seed (achene): leaf-blades varying widely from deeply cut into linear segments *(C. verticillata)* to the lance-shaped kind *(C. lanceolata)*, mostly opposite. One species *(C. rosea)* has pinkish purple rays, the remainder are yellow. In a wide range of soil habitats. Summer.

474. Beggar-ticks *(Bidens).* Medium to tall annuals or biennials, 2-6 ft. high, with one or few yellow-rayed heads (rays sometimes absent) borne at the branch ends; bracts in 2 series, the outer longer than the inner; calyx of 2-4 slender, barbed teeth persistent on the seeds (achenes): leaves both simple and compound, opposite, the blades with toothed margins. About 9 species distributed in moist soil and marshes. Summer.

475. Loud speakers (*Marshallia,* fig. 208). Perennials, 6-28 in. tall with solitary long-stalked rayless heads borne terminating the branch ends; flowers all alike, the purplish to white corollas funnel form radiating from the receptacle like a cluster of loud speakers; bracts of equal length in 1 or 2 series; seeds 5-angled: leaves alternate, the blades linear to ovate, mostly with 3 principal veins. About 6 species distributed in open woodlands and the eastern savannah lands. Summer.

476. Sneezeweed *(Helenium).* Mostly perennial herbs, 8 in.-6 ft. tall, with showy yellow-rayed heads in a spreading, open inflorescence above; central yellow or purplish flower mass hemispherical in shape with no chaffy scales among the flowers; bracts of head spreading or even reflexed; calyx of 5-8 bristle-pointed scales: leaves alternate, the blades linear, or lanceolate to ovate, the broader blades commonly toothed. In moist soil of open areas and marsh borders. Summer.

The fine-leaved common bitterweed *(H. tenuifolium)* of old fields and roadsides belongs here. Especially attractive are the tall marsh sneezeweed *(H. autumnale)* with yellow central flowers, and the purple-head sneezeweed of moist soil.

477. Gaillardia *(Gaillardia lanceolata).* Annual or biennial, 1-2 ft. high, widely branching bearing scattered long-stalked purple and yellow-rayed heads; the rays on some plants are merely enlarged tubular flowers; central flower mass purple; calyx of 6-9 bristle-like scales longer than the seeds; rays tipped with prominent teeth: leaf-blades lance-shaped or elliptical, mostly sessile, finely pubescent. In dry sandy soil. Adventive from the western United States. Abundant in and around Southport. Spring and summer.

478. Northern dog-fennel *(Anthemis Cotula).* Annual, strong-scented plants, 1-2 ft. tall, with the heads of white-

FIG. 209. MAKING A DAISY CHAIN WOULD BE AN EASY TASK HERE. PHOTO BY H. L. BLOMQUIST.

rayed flowers terminating the numerous branches; central flowers yellow in a conical mass: leaf-blades finely cut into thread-like segments, alternate. A weed from Europe, chiefly in the western region.

479. Yarrow *(Achillea Millefolium)*. Perennial weed, 1-2 ft. tall, with a flat-topped cluster of small heads with 4-6 white-rayed (sometimes pink) flowers: leaf-blades very finely cut or dissected, the basal ones stalked. In waste places but not common eastward; a plant immigrant from Europe. Summer and fall.

480. Field daisy (*Chrysanthemum leucanthemum*, fig. 209). Frequent perennial weed 1-3 ft. high with showy white-rayed heads borne terminally on long stalks; stem leaves mostly narrow and coarsely toothed. Summer and fall. In old fields but not common eastward; a plant immigrant from Europe and Asia.

This foreign weed has been reported as the state flower. This is incorrect for when North Carolina chooses a ''state flower'' it will be a native North Carolina plant.

481. **Leopard's-bane** *(Arnica acaulis).* Erect single stemmed perennial bearing an umbel-like cluster of stalked yellow-rayed heads; rays nearly 1 in. long; central flower mass flat, without chaffy scales among the flowers: basal leaves 4, broad, lying flat on the ground, stem leaves small, 1-3 pairs, leaves and stem hairy. In moist woods. Spring.

482. **Composite fireweed** *(Erechtites hieracifolia).* Coarse annual, 2-8 ft. high, the branches terminated by a loose cluster of green heads since the corollas barely show beyond the bracts which enclose the head in a single series; no ray flowers; no chaffy scales among the flowers; base of head swollen: leaves alternate, blades ovate to lance-shaped, coarsely toothed or sometimes deeply cut. In woods, especially common following fire. Summer.

483. **Indian plantain** *(Mesadenia).* Tall perennials with milky juice, 3-10 ft. high, with a large terminal loose cluster of narrow or cylinder-shaped white or yellowish heads; bracts 5, of equal length in one series; corollas all alike; calyx of fine bristles: leaves large at base, smaller above, blades mostly of the broad orbicular or ovate type, lobed in one species. In moist soils of woods and open bog lands. Summer.

484. **Sweet-scented Indian plantain** *(Synosma suaveolens).* Similar to the preceding but not over 5 ft. tall; bracts of the heads 12-15 enclosing a few flowers all alike with no chaffy scales: leaf-blades triangular in outline, the lower lobed with a heart-shaped base, the upper arrow-head shaped, toothed. In woods especially in the mountains. Late summer.

485. **Ragwort** *(Senecio).* Mostly perennial herbs, 1-3 ft. tall, with terminal loose clusters of rather showy yellow-rayed heads; bracts of head about the same length in one series; no chaffy scales among the flowers; central flowers in a nearly flat-topped mass; calyx of white bristles: leaves varying in our 11 species from simple to deeply lobed, chiefly basal. In the very showy golden ragwort of the western area the leaf-blades are ovate with a heart-shaped base and in the common

old field and open woodland species of the middle and eastern region, the leaf-blades are narrow lance-shaped tapering into slender stalks. Both have small rounded teeth. In wet and dry soils. Summer.

486. Thistle (*Carduus*, fig. 65). Coarse biennials or perennials, 2-9 ft. high, with large bristly heads borne singly or clustered; bracts spine-tipped; corollas all tubular; calyx of smooth or plume-like bristles: leaves in our 7 species ranging from simple to deeply lobed and cut, all more or less spiny. In upland open areas. Summer and fall.

487. Night-nodding, bog dandelion *(Thyrsanthema semifloscuare).* Low perennial with flower stalks, 3-7 in. high, bearing single terminal heads which are nearly erect in the sunshine but nodding on dark days and at night; central flowers slightly 2-lipped; ray flowers white within purplish without: leaf-blades in a basal rosette, long-elliptic, smooth above, woolly beneath. In grassy bogs or savannahs of the coastal plain. Early spring.

488. White lettuce or rattlesnake-root *(Nabalus).* About 8 species of medium to tall perennial herbs ranging from 2-7 ft. in height, bearing loose terminal clusters of nodding or mostly white drooping slender heads; heads narrow or cylindric with a few flowers enclosed in 1 or 2 series of equal bracts light in color; flowers all of the ray-type (corolla strap-shaped) the rays not spreading: leaves simple or deeply lobed, alternate on the erect stems, the upper ones commonly clasping the stem. Chiefly in woods but one species *(N. virgatus)* in savannahs.

489. Hawkweed *(Hieracium).* About 6 species of medium perennials ranging from 1-3 ft. tall, the few-leaved stem branching above into a loose inflorescence of yellow-flowered heads; flowers all of the ray-type (corollas strap-shaped), spreading; bracts overlapping, the outer reduced; calyx of 1 or 2 rows of brown bristles: leaves more or less hairy, chiefly basal or largest at the base, blades lance-shaped or nearly so, shallowly toothed or entire. In dry woods. Late spring and summer.

490. False dandelion *(Sitilias Caroliniana)*. Biennials or perennials, 2-5 ft. high, bearing one or few very showy yellow-flowered heads on the branch ends; flowers all of the ray type, each ray 5-toothed at the end; bracts about equal in one series; calyx of brown bristles with a ring of white hairs at the base: leaves alternate, blades lance-shaped in outline, the lower deeply lobed, the upper with a few coarse teeth, smooth. In dry soil of open areas. Spring and summer.

491. Wild lettuce *(Lactuca)*. About 7 species of mostly tall biennials, ranging 2-12 ft. in height, bearing terminal open or very loose clusters of white, yellow or blue-rayed heads; all flowers of the ray-type, the rays 5-toothed at end, body of head straight-sided, the inner row of bracts equal in length, the outer bracts shorter: leaves simple, blades ovate, or narrower in outline, mostly toothed and deeply lobed in some species, mostly smooth. In various soils of open-areas. Summer.

492. Dandelion *(Taraxacum Taraxacum)*. Perennial weed with the yellow flowers in a single head at end of hollow stem; calyx (pappus) in fruit borne on a slender stalk high above the ovary. A weed found in all continents. Spring and summer. Leaves coarsely toothed and cut, forming basal rosette.

Most botanists recognize in the dandelion the most highly specialized plant in the world. One of its peculiarities is that it has given up reproduction by pollination and fertilization; the seeds develop normally without these processes.

493. Salsify *(Tragopogon porrifolius)*. Coarse perennial, 2-5 ft. tall, with showy heads (over 2 in. broad) of bluish purple flowers which close by the middle of the day: leaf-blade narrow, tapering to a sharp pointed top, the base clasping the stem. Occasional in waste places; a weed from Europe. Summer and fall.

494. Dwarf false dandelion *(Adopogon).* Four species of stemless annuals or perennials ranging 6-30 in. high, bearing commonly but a single orange or yellow-rayed head; flowers all of the ray type; bracts narrow in 1 or 2 rows, those in each row equal; calyx a row of bristles with a row of scales commonly present without: leaf-blades basal, lance-shaped in outline, few-toothed or lobed, sessile or tapering into short stalks, smooth. Mostly in moist soils of open upland areas. Spring and summer.

495. Chicory *(Cichorium Intybus).* Attractive perennial weed, 1-5 ft. tall, with sky-blue or sometimes white heads widely scattered on the branch ends; heads open in the morning closing by midday; bracts surrounding heads in two series, the outer spreading: leaf-blades variable, the margins coarsely toothed and cut. A weed from Europe widely scattered in waste places.

INDEX TO SCIENTIFIC NAMES

NOTE: The herbaceous plants are listed chiefly by the genus name only. For the species treated under such genera, refer to the manual (Section II) by the page number given.

INDEX TO SCIENTIFIC NAMES

INDEX TO COMMON NAMES

INDEX TO COMMON NAMES